FROM SMOKE SIGNALS TO SATELLITES

THE FASCINATING STORY OF COMMUNICATION

D.K. DAS

Former Director and Distinguished Scientist
Space Applications Centre, ISRO, Ahmedabad

P.S. BHARADHWAJ

Former Senior Scientist
Space Applications Centre, ISRO, Ahmedabad

ISBN 979-8-89632-301-3

The authors dedicate this book to all the students of India who dream of becoming space scientists one day.

Contents

एन एम देसाई / *N M Desai*
विशिष्ट वैज्ञानिक / *Distinguished Scientist*
निदेशक / *Director*

सत्यमेव जयते

भारत सरकार GOVERNMENT OF INDIA
अंतरिक्ष विभाग DEPARTMENT OF SPACE
अंतरिक्ष उपयोग केंद्र
SPACE APPLICATIONS CENTRE
अहमदाबाद AHMEDABAD - 380015
(भारत) / (INDIA)
दूरभाष / PHONE:+91-79-26913344, 26928401
फैक्स / FAX : +91-79-26915843
ई-मेल / E-mail: director@sac.isro.gov.in

Foreword

It is with great pleasure that I *introduce this book "From Smoke Signals to Satellites," authored by Shri D K Das and Shri P S Bharadhwaj, both Satellite Communication veterans of the Indian Space Research Organization (ISRO).* Having dedicated their careers to the Space Applications Centre (SAC), they have significantly influenced the landscape of satellite communications in India.

This book not only transcends historical account of communications technology but it tells the inspiring story of human aspirations, innovation, and collaboration that have paved the way for a connected world. It traces the journey of communication systems from their humble beginnings to the cutting-edge satellite technologies that are integral to modern life today highlighting ISRO's pivotal role in this transformation and its relentless pursuit of technological excellence.

One of the USP features of this book is its accessibility to a wider audience. The authors have skillfully presented complex technical concepts in an engaging and reader-friendly manner, making this valuable knowledge available to everyone, regardless of their background or expertise.

I am confident that this book will resonate with many readers, especially as ISRO's achievements - such as missions to the moon and beyond have captured the imagination of millions worldwide. "From Smoke Signals to Satellites" enriches this narrative by providing a thorough perspective on ISRO's contributions to the communications revolution.

I would like to extend my heartfelt congratulations to the authors Shri D K Das and Shri P S Bharadhwaj, for their remarkable contributions in the field of communications and for bringing this enlightening book to life. I also thank everyone who supported and nurtured the book from its initial concept to publication.

I invite you to explore the stories of discovery and innovation within these pages. I believe that in your journey through the milestones of communications technology, you will find inspiration from many individuals featured in this narrative.

Place: Ahmedabad

(एन एम देसाई) / (N M Desai)
निदेशक / Director

भारतीय अंतरिक्ष अनुसंधान संगठन  **INDIAN SPACE RESEARCH ORGANISATION**

Preface

In this book, we delve into the fascinating history of Communication and explore how it has shaped the world we live in today. From the early days of smoke signals and carrier pigeons to the advanced satellite communications of today, we will follow the stories of the pioneers who pushed the boundaries of what was possible and changed the way we communicate forever.

Through vivid descriptions and compelling human narratives, we will uncover the key moments in the history of Communication and the people who made them happen. We will also delve into the rich history of modern telecoms in India, including the inspiring story of ISRO and the Indian Space Programme.

This book is not just about the technical details of communication innovations, but also about the human impact they have had on our world. So join us on a journey through time as we explore the evolution of communication and the people who made it possible.

Sincerely,
D K Das
P S Bharadhwaj

Acknowledgements

We would like to express our heartfelt gratitude to the following individuals, whose support and guidance were instrumental in making this book possible.

First and foremost, we extend our deepest thanks to Shri Nilesh Desai, Director, Space Applications Centre (SAC), ISRO, for his unwavering support and for providing SAC resources throughout the writing process.

We are profoundly grateful to Shri D. K. Singh, Director, Human Space Flight Centre (HSFC), ISRO, for his invaluable contributions to the concept of this book.

Special thanks to Dr. Abhijit Chatterji, former Scientist, SAC, ISRO, for his exceptional work on the cover page and illustrations, which have greatly enhanced the quality of this book.

Our sincere appreciation goes to Dr. S. C. Bera, Deputy Director, SAC, ISRO, as well as Shri Rajesh Kumar Singh and Dr. Mehul Pandya, Senior Scientists, SAC, ISRO, for their insightful feedback and meticulous review of the draft, reading it word by word to improve its content.

We also extend our gratitude to Dr. Abha Chhabra and Shri Sumit Kumar, Senior Scientists, SAC, ISRO, whose initiative and dedication were instrumental in bringing this book to life. Their efforts in connecting with the publisher, coordinating throughout the process, and managing the editing and formatting were crucial to the book's success.

Finally, we thank M/S Notion Press for providing the platform and support needed to publish this book.

To each and every one of you, we offer our deepest thanks for your unwavering support and encouragement.

Part-1
Terrestrial Communication

1

The Beginning

Communication is the solvent of all problems and is the foundation for personal development.

- Peter Shepherd

Since prehistoric times, staying in touch with one another has been a major human need. This is reflected in mythology and literature from all over the world.

The *Ramayana* tells us of Hanuman conveying messages on behalf of Lord Ram. In *Mahabharata*, Krishna goes to the Kaurava's Court as a messenger of the Pandavas and gives Duryodhana an ultimatum - either surrender the disputed land or perish. Also from *Mahabharata* is the romantic legend about a swan

Figure 1: *Hamsa Damyanti by Raja Ravi Verma (Image: Public Domain)*

15

being used to convey the message of love from prince Nal to Princess Damayanti. The *Atharvaveda* refers to couriers by the term 'palagala'.

Figure 2: Meghdoot by Ram Gopal Vijaivargiya
(Image Source: Maharaja Sawai Man Singh II Museum)

Meghdoot or *The Cloud Messenger,* a 4th century classical Sanskrit masterpiece by Kalidas, has enchanted people over the generations. It is a tale of separation and longing of lovers during the romantic monsoon season. Unable to bear the separation, a Yaksha (celestial attendant) requests a passing cloud to carry a message to his young bride. Kalidas was a court-poet; yet in Meghdoot, his description of nature and the Yaksha's plight touches a universal chord. His narrative sparks the imagination and enables the reader to experience cloud's journey.

The importance of communications is likewise reflected in Greek and Roman mythology: Hermes and Mercury were the presiding deities over matters like swiftness, travel, trade and the sending of messages. From the Old Testament we have Noah using carrier pigeons.

In practical terms, certain basic means of communication have predated the advent of written language. These include smoke signals and drums, carrier pigeons, runners etc. Evolution was extremely slow for the first few millennia, so that communications methods from the 1700s would perhaps offer no surprises to a Greek, Roman or Harappan citizen.

Figure 3: Noah and the Dove, Mosaic in Basilica di San Marco, Venice (Image Source: Public Domain)

The early stages of the scientific and industrial revolution brought in more efficient and better organized applications of the old ideas. For example, a semaphore network was set up spanning large parts of France, using signal stations similar to railway signals. These were installed on hill tops, and a message could be relayed from hill to hill by trained operators, covering hundreds of kilometers in a couple of hours. The same result could be achieved by the *heliograph,* an instrument that could send messages by means of sunlight reflected from a mirror.

It was not until the 1830s that a few pioneers began to apply the brand new science of electricity and magnetism and come up with the revolutionary communications techniques that brought us today's cable, fiber and wireless systems. This book presents the fascinating story of man's long journey from smoke signals to modern-day satellite communications and mobile technology. We will be telling this story from a global as well as an Indian perspective.

2

Smoke Signals

We had a smoke signal from the other side, but unfortunately it was all white.

– Eugene Cernan

One of the first attested examples of long-distance signaling dates back to around 1800 BC in China. Soldiers patrolling the Great Wall used smoke signals to warn their comrades hundreds of miles away when invaders appeared on the horizon. They even used color-coded smoke to add extra information such as the size of the invading force. They used a mixture of wolf dung, saltpeter and sulfur to create dense plumes of smoke that could be seen from a distant watch tower and relayed to the nearest headquarters.

A literary Chinese idiom dating from around the 9th century, "wolf smoke rising around us", was often used to describe the looming threat of war.

Some Native American tribes used smoke signals to send messages over long distances. Each tribe had their own code to convey different meanings. For example, if smoke came from halfway up the hill, this would mean that all was well, but from the top of the hill it would be a danger warning.

In ancient Sri Lanka soldiers stationed on the mountain peaks would alert each other of impending attack from British, Dutch and Portuguese invaders by signaling from peak to peak. In this way, they were able to transmit a message to the King in just a few hours.

Figure 4: An artist's view of smoke signals on the Great Wall Watch Towers by Dr. Abhijit Chatterjee

A ceremonial remnant of the old days continues today in the Vatican, where the College of Cardinals uses smoke signals to announce the selection of a new Pope during a papal conclave. Eligible cardinals conduct a secret ballot until one candidate receives a vote of two-thirds plus one. The ballots are burned after each vote. Crowds of people assemble outside the building and watch eagerly for the smoke to rise from the chimney. If the smoke is white, a huge cheer goes up, for it means that a new Pope has been elected. Black smoke means that no candidate was elected in that round, and the people outside must settle down for a longer wait.

Figure 5: (Image Source: Mental Floss, Mar 13, 2013)

Smoke signals are still used in modern combat situations, using special smoke grenades that produce different colours of smoke. For example, a unit on the ground might use green smoke to let a supporting aircraft know their location and thus avoid a mistaken attack on their own position. Other colours could be used for specific purposes like requesting evacuation by helicopter. This could be specially useful when the ground personnel are hidden under a forest canopy.

Figure 6: Army Special Forces soldiers use a smoke grenade to signal a UH-60 Black Hawk medical evacuation helicopter for a medical airlift during Southern Strike 17 at the Gulfport Combat Readiness Training Center, Miss., Oct. 26, 2017. Air Force photo by Airman 1st Class Sean Carnes. (Image Source: Public Domain)

3

Signal Drums

A people without the knowledge of their past history, origin, and culture is like a tree without roots.

– Marcus Garvey

A joyful group of people circles a campfire, chanting and dancing enthusiastically to the beat of a jungle drum. But woven into the mesmerizing rhythm, is there perhaps a completely different message for distant ears?

A talking drum is a type of drum that, by imitating the rhythm and the rise and fall of words in languages, are used as communication devices. In Africa, New Guinea and tropical America, people have used drums to communicate with each other for centuries. When European expeditions arrived, they were surprised to find that news of their doings and intentions was carried through the woods faster than they could ever march. African drummers can transmit a message at the speed of 100 miles per hour.

Among the famous communication drums are those of West Africa. From present-day Nigeria and Ghana they spread across West Africa and to America and the Caribbean during the slave trade. There such drums were banned because they were being used by the slaves to communicate over long distances in a code unknown to the slave traders.

Drums were used in European conflicts as well, such as during the English Civil War when drummers were employed to relay commands over the noise of battle. Different units were differentiated from each other by using distinct drum beats to avoid confusion. Alternatively, bugles and trumpets were also used. This practice evolved into military bands with distinctive features such as the bagpipes of the Scottish Highland regiments.

Figure 7: *Slit Drum. (Image Source: Wikipedia, Public Domain)*

Central American civilizations like the Aztecs also used drums to communicate during warfare.

War drums called "Dundhubi" figure extensively in narratives of war in Vedic scriptures. Book VI of the Rig Veda contains a "Hymn to the Battle Drum", evoking the image of Aryans charging into battle to the beat of drums. [24]

4

Pigeon Post

In the days before modern inventions, the sky was owned by pigeons – the messengers of a thousand journeys, bearers of news across time and space.

- H.G. Wells

Pigeons have carried messages for thousands of years and can deliver messages much more quickly than people. They were even used by pharaohs to deliver messages in Ancient Egypt. Pigeons were reliable for delivering messages because they always found their way home.

Figure 8: (Image Credit: BenTheWikiMan, Public Domain)

The homing pigeon, also called the mail pigeon or carrier pigeon, is a variety of domestic pigeon which has the ability to find its way home over extremely long distances. Because of this skill, such domesticated pigeons were used to carry messages.

They were transported to a destination in cages, where they would be attached with messages, then naturally the pigeon would fly back to its home where the owner could read his mail.

They have been used in many places around the world. In World War II, troops used pigeons to send messages to each other. Pigeons were very useful in times of war because they could fly above the battlefield and through smoke. The Dickin Medal, the highest possible decoration for valor given to animals in UK, was awarded to 32 pigeons for their great service during the World War II.

Figure 9: Young lady in oriental clothing with a homing pigeon (19th century painting) Image Source: Wikipedia, Public Domain

Cher Ami: Flight of Courage

An iconic example of a pigeon's historic role is the tale of Cher Ami ("Dear Friend" in French), a pigeon born (or rather hatched) in 1918. Cher Ami was a member of a 600-strong "unit" of Army Signal Corps pigeons that provided vital communication on the battlefields of World War I. At that early stage of electrical communications technology, pigeons were still crucial for tactical communications.

Cher Ami flew 12 successful missions, which was a fairly good track record for army pigeons. Once the Germans realized what was going on, they began to shoot down any suspected "spy birds". Cher Ami completed his most perilous mission during the Meuse-Argonne offensive in 1918, when he was stationed with the 77th Division of the U.S. Army, – the so-called Lost Battalion in the Argonne Forest. Cut off from the rest of the American troops, and

Figure 10: *The stuffed body of Cher Ami on display at the Smithsonian Institution. (Image Source: Public Domain)*

facing heavy bombardment, the battalion was desperate to signal their plight to their comrades. To make matters worse, they came under "friendly fire" and bombardment from American aircraft on October 4.

They dispatched many pigeons with distress messages. Maj. Charles Whittlesey and his men watched in despair as pigeon after pigeon was brought down by enemy fire. Cher Ami was the last surviving bird. Whittlesey attached a note to his leg, with this message: "We are along the road parallel to 276.4. Our own artillery is dropping a barrage directly on us. For heaven's sake, stop it."

Cher Ami dodged a hail of bullets, giving the soldiers a ray of hope – until he too was shot and fluttered towards the ground. Despite his injury, he returned to the sky and completed his perilous 25 mile journey in record setting half an hour. Army medics worked hard and barely saved Cher Ami's life. The bird survived the war with one leg and no eyesight.

As soon as the message was received, the unintentional bombardment on the Lost Battalion was called off and the

194 men safely returned to American lines. Cher Ami received the Croix de Guerre from the French government for his gallant and crucial mission. Under the care of his trainer, Capt. John Carney, Cher Ami returned to the United States where he spent the rest of his life. Cher Ami died of his war wounds on June 13, 1919, in Fort Monmouth, New Jersey. His body was preserved and presented to the Smithsonian Institution with honor in 1921. [22]

G.I. Joe: In the Nick of Time

A somewhat similar situation occurred in October 1943 during World War II, when an American bombing raid was planned on the German-occupied town of Colvi Vecchia in Italy. Just before the scheduled air raid, the British 56th Infantry Brigade was able to overcome the German defences and capture the town, forcing the Germans to retreat. Now the British officers had to send an urgent message asking that the bombing be called off. Failing to make radio contact, they had to fall back on their carrier pigeons. A bird named "G.I. Joe" was despatched, flying 20 miles in 20 minutes. The bombers were already lining up for takeoff when G.I. Joe landed with the vital message and saved the 150 British troops who were holding the town. The British awarded G.I. Joe the Dickin Medal for Gallantry. [23]

In India, Chandragupta Maurya, Emperor Ashok and the Mughal Emperors used pigeons as message carriers.

Figure 11: G.I. Joe with his Dickin Medal

The last pigeon messaging service in the world was the Orissa Police Carrier Pigeon Service India. In 1946 the Orissa Police Department took over 40 pigeons from the departing colonial government at the close of the Second World War. The service began in the mountainous Koraput district and was soon expanded to cover 38 places. During its heyday the Carrier Pigeon Service boasted of a 1,500-strong trained fleet. The service was so useful that it was continued until 2008, and in fact exists as a ceremonial unit to this day.

Figure 12: Image Credit: Official Web Portal of Odisha Police

In April 1948, when Jawaharlal Nehru had visited Odisha, he desired certain changes in the public meeting scheduled at Cuttack. A homing pigeon of the State Police transported his message from Sambalpur to Cuttack 400 km. away in hours, to his great delight.

During the Postal Centenary celebrations of 1954 at Delhi the Homing Pigeons of the Orissa police gave a successful demonstration, by conveying the message of inauguration from the President to the Prime Minister.

In the 1980s, carrier pigeons used to fly to remote areas affected by insurgency and where the wireless network had limited range. They were very effective.

In 1982, when massive floods hit coastal Odisha, disrupting road connectivity for weeks, pigeons were the ones delivering messages. They were so much an integral part of the police administration that till 2010 it was mandatory for newly-recruited officers to clear a 10-mark test on pigeon service

Figure 13: *Image Credit: Storypic, 16th April 2018*

However, in 2000, the Comptroller and Auditor General of India had expressed doubts about its relevance, deeming it a "wasteful expenditure" and recommending that it be wound up. In response, intellectuals, conservationists and traditionalists in Odisha protested and demanded that at least a skeletal service be maintained for its heritage value. Although operational use ceased in March 2008, about 150 pigeons continue to be maintained for ceremonial purposes in Cuttack at the Police Training College in Angul. The service costs about ₹1 lakh each year, and the sub-inspector and two constables employed in the service draw salaries from the government like any other police personnel. [25]

5

Runner

To send a letter is a good way to go somewhere without moving anything but your heart

– Phyllis Theroux

The Battle of Marathon was a turning point in history, when the mighty Persian empire, already ancient at that time, lost to the rising power of the Greek city-states in 490 BC. The Greeks wanted to make sure of their victory and prepare against a fresh counter-attack on Athens. The task of warning that city was entrusted to a runner named Pheidippides or Philippides. He ran a Marathon race against time and sounded the alert, covering 40 kilometers in a couple of hours, only to die of exhaustion soon after.

Figure 14: Statue of Pheidippides. (Image Source: Wikipedia, Public Domain)

In India, one of the earliest evidence of a systematic postal service using foot messengers is found during the reign of Chandragupta Maurya (322-298 B.C.). Emperor Ashoka also devised a very efficient means of communication of mail runners, horse couriers, pigeon carriers and camels for official communication

When Ibn Batuta was travelling in India, in the middle of the 14th century, he found an organised system of couriers established throughout the country by Mohammed Bin Tughlak. This system was of course established for the convenience of the Emperor and was continued with various innovations by successive Moghul emperors until the 18th century.

Figure 15: Dawk Wallahs of Bengal, 1858.
(Image Source: The Illustrated London News, 20 February 1858.)

The East India Company established a postal system of its own to facilitate the conveyance of letters between different offices;

but it was only in 1774, during Warren Hastings' administration that a Postmaster General was appointed and the general public could avail of the service, paying a fee on their letter.

During the British Raj, each runner would cover 12-13 km to exchange the postbag with another runner. Some ran nearly 20 km a day. In the process, they encountered wild animals, bandits and risked their lives. There were cases when the runners on duty were carried away by tigers, drowned in flooded rivers, bitten by venomous snakes, buried in avalanche or murdered by robbers. In the face of all these dangers, the Runners seldom shrunk from their responsibilities. The runners used to put on colorful attire with badges and were armed with spears and jingling bells for self-defense.

Mail-runners faded with the advent of railways in the late 19th century, but continued to work in far-flung areas. They remain the only means of communication in remote regions. They may not call themselves mail-runners, but some postmen in India still deliver letters in hilly areas on foot.

Rudyard Kipling wrote this tribute to the mail runners in the Govenment's service in British India:

Figure 16: Image source: Solvyns, (1799), Sec. II, No. 20.

The Overland Mail

In the name of the Empress of India, make way,
O Lords of the Jungle, wherever you roam.
The woods are astir at the close of the day—
We exiles are waiting for letters from Home.
Let the robber retreat—let the tiger turn tail—
In the Name of the Empress, the Overland Mail!

With a jingle of bells as the dusk gathers in,
He turns to the foot-path that heads up the hill—
The bags on his back and a cloth round his chin,
And, tucked in his waist-belt, the Post Office bill:—
"Despatched on this date, as received by the rail,
"Per runner, two bags of the Overland Mail."

Is the torrent in spate? He must ford it or swim.
Has the rain wrecked the road? He must climb by the cliff.
Does the tempest cry halt? What are tempests to him?
The Service admits not a "but" or and "if."
While the breath's in his mouth, he must bear without fail,
In the Name of the Empress, the Overland Mail.

From aloe to rose-oak, from rose-oak to fir,
From level to upland, from upland to crest,
From rice-field to rock-ridge, from rock-ridge to spur,
Fly the soft sandalled feet, strains the brawny brown chest.
From rail to ravine—to the peak from the vale—
Up, up through the night goes the Overland Mail.

There's a speck on the hillside, a dot on the road—
A jingle of bells on the foot-path below—
There's a scuffle above in the monkey's abode—
The world is awake, and the clouds are aglow.
For the great Sun himself must attend to the hail:—
"In the name of the Empress, the Overland Mail!"

– Rudyard Kipling
(1886)

The Unknown Post Man and the National Anthem of Bangladesh

He was Gaganchandra Dam (1845–1910), mostly known as Gagan Harkara, a postman at Shelaidaha Post Office in Kumarkhali in present-day Bangladesh. He was also a Baul lyricist. Gagan used to visit Rabindranath quite frequently to deliver letters, and used to have had discussions about Baul. Rabindranath used to listen to Gagan's songs on the roof of a boat, and this had a far-reaching impact on Tagore's mind and after the tune of Gagan's famous song "Ami Kothay Pabo Tare", Rabindranath Tagore composed "Amar Shonar Bangla", the National anthem of Bangladesh.[15]

Figure 17: A statue of Gagan Harkara in Kusthia, Bangladesh (Image: Prothom Alo Bangla 24 Mar 2017)

6

Telegraph

The invention of the telegraph is to civilization what the nervous system is to the human body.

- Arthur C. Clarke

Samuel Morse and the Telegraph: Tragic Beginnings

Two teams of inventors simultaneously developed the electrical telegraph: Wheatstone and Cooke in England, and Samuel Morse in the United States.

Morse took up the challenge of electrical telegraphy after his wife Lucretia had a heart attack during childbirth in New Haven, while Morse was away in Washington DC. After receiving a letter from his father, he rushed back immediately. Sadly, Lucretia had died and was already buried by the time he could arrive. Heartbroken, Morse resolved that he would find a way to send messages instantly over long distances, so that others would be spared the pain of such situations.[10]

Eight years later, he met Samuel Charles Thomas Jackson, who shared his interest in science and technology in general and electrical communication in particular. These conversations rekindled Morse's keen interest in telegraphy, and he began sketching out preliminary designs for the instrument.

Figure 18: *Samuel Morse by uthoMathew Benjamin Brady (Image: Public Domain)*

Morse decided to call his device the telegraph. He created a code that used long and short pulses (called dots and dashes) to represent the alphabet, numerals and punctuation marks. He then began working on a system that could transmit such codes over wires. Nearly a decade later, he was ready to demonstrate his invention to the public at an event that he organized at the Speedwell Ironworks Factory. On January 11, 1838, Samuel Morse successfully used his code of dots and dashes to transmit the message, **"A patient waiter is no loser"**

Thirteen years after his beloved wife died, Samuel Morse's innovation changed the way the world communicated. In the process, it gave people an opportunity to ensure that they and their loved ones were connected for the important moments in their lives.

At first, the received dots and dashes were recorded mechanically on a moving paper tape and decoded later. But it was soon found that an

Figure 19: *Morse key. Image Credit: Britannica*

experienced operator could listen to the sound of the instrument and decode the message in his mind. This method became the standard and the tape system was discarded.

An Inventor's Marriage Proposal

Although a tragic death may have led to the invention of the telegraph, the Morse code figures in a happier story whose hero is none other than Thomas Edison.

Edison had begun his career as a railway telegraph operator. After meeting his second-wife-to-be Mina Miller, he taught her Morse code so that they could secretly communicate with each other while her family was around.

And when he decided to propose to her, he signaled in Morse code, "Will you marry me?" to which she replied "Yes."

If this doesn't show that Edison was passionate about Morse code, consider this: Edison's first two children were nicknamed Dot and Dash. If that is not enough, here's another fun fact: Edison had some Morse code tattooed on his arm (probably during his youth as an apprentice telegrapher).[16]

Figure 20: An Inventor's Marriage Proposal, Representing Image

The Telegraph Comes to India

In India, an assistant surgeon with the East India Company named William O'Shaughnessy began independently to develop an electric telegraph system. In 1839, he set up a 13½-mile-long demonstration line near Calcutta. That was only two years after Samuel Morse built his system in the United States, although O'Shaughnessy was unaware of Morse's work.

Figure 21: Sir William Brooke O'Shaughnessy, MD FRS. The father of Indian Telegraphy (Image: Public Domain)

His first version did not have an electromagnetic receiver. Instead, the operator would place a finger on a metal disc and sense the mild electric shocks from the incoming electric signal, while writing out the message with the other hand. (No, this is not the origin of the expression "shocking news").

O'shaughnessy, however, could not get backing for commercial telegraphy in India until 1847. Lord Dalhousie, the new Governor General of India, saw the potential of the invention and authorized him to build a 27-mile line between Calcutta and Diamond Harbour. By 1851 the usefulness of the system was so well proven that Dalhousie authorized an All-India telegraph network. O'Shaughnessy built up the network over the next three years, and in 1854, the service was opened to the public. The first telegram was sent from Mumbai to Pune in that year. The telegraph system proved vital for the British to suppress the revolt of 1857. They were able to mobilize their troops swiftly and checkmate the rebel sepoys' moves at every stage. The story goes that one rebel prisoner, while he was being led to the gallows, pointed to a telegraph line and declared: **"There is the accursed string that strangles us"** [17]

Figure 22: "There is the accursed string that strangles us"
An artist's view by Dr. Abhijit Chatterjee

In the mid-20th century, tele printers (TELEX) and facsimile (FAX) superseded the manual key telegraph, and in turn they were obsoleted by electronic messaging over computer networks. But in India, the manual telegraph continued into the late 20th century and beyond. The 1980s were the golden years of the Telegraph service in India, with more than 100,000 telegrams being sent and received per day via the Delhi main office alone.

The telephony and Internet revolution in the late 1990s and early 2000s finally doomed the telegraph service to oblivion. By 2011, it was a colossal loss-making operation, which was finally closed on 15th July, 2013. On the last day of its 160 year history, thousands of telegrams were sent by people as part of a commemorative event.

Figure 23: An employee of the National Telecom Museum in Bhopal, Madhya Pradesh, sitting behind an 1871 model of a telegraphic code machine.
Image Credit: Sanjeev Gupta/European Press photo Agency

The world's very last telegram was sent by one Ashwani Mishra, according to a report from PTI. Mishra managed to make history of sorts by sending the very last telegraph message just before closing time at 11:45 PM on the last day of the service. He used this opportunity to send a telegram to Shri Rahul Gandhi from the Central Telegraph Office (CTO), Janpath. He also sent a message to the then Director General of Doordarshan [18]

7

Telephones

I think it's the most amazing invention that has come along in my lifetime–The telephone. It changed everything

- George Carlin

"**Mr. Watson, come here, I want to see you.**" This sentence entered the history books in 1876, when Alexander Graham Bell and his assistant were working in two separate rooms, with an experimental telephone line running between them. Bell made the first successful test of the instrument by trying to summon Mr. Watson by speaking into the prototype microphone. To his delight, his assistant soon rushed into the room and confirmed that he had heard Bell's voice clearly through the earpiece.

An early ancestor of the modern telephone can be seen in the acoustic string telephone - you can make one for yourself with a length of string and two paper cups. Some big offices used "speaking tubes" that enabled people to talk from room to

Figure 24: Telephone-Alexander Graham Bell (1876). Image Credit: Sutori

room. They were commonly used between the bridge and the engine room of a ship.

Around the mid 1800s, many people were trying to transmit sound using electromagnetic instruments. History credits Bell with the first success, as he filed a patent in 1876 for an "apparatus for transmitting vocal or other sounds telegraphically".

The first telephones were directly connected to each other from one customer's office or residence to another customer's location. A group of customers located near to each other would often be connected together on a "party line". If you wanted to call someone else on the same party line, you would turn a handle on the instrument, ringing a bell on each of the other party phones. Each member of the group would have a code like "long-short-long". On hearing their code, the called party would answer the call while others would ignore the ring.

Soon, manual switching exchanges were set up to connect subscribers to each other on a one-to-one basis, or to switch calls to other faraway exchanges. As the demand for connections grew, automatic exchanges were developed, and networks quickly grew in size and complexity.

The early history of telephones is full of strange and interesting stages of development. For example, in some rural areas of the United States, it was often uneconomical to install the usual copper wires to remote, isolated farm houses. Instead, existing barbed wire fences were sometimes used to connect local farms together into a crude telephone service.

Why Do We Say Hello When Answering The Phone?

Though we credit Alexander Graham Bell as the inventor of the telephone, it was none other than Thomas Edison who coined the telephone greeting of "hello", which is today understood in every corner of the world.

Figure 25: (Image Credit: Adam Cole, NPR, 17/2/2011)

The very first telephone lines were used to link two instruments, or at most three or four instruments that would be "always on". No bells were installed at that stage of development, so you would have to literally "call" out loudly over the phone, hoping that someone at the other end would hear your voice and reply. In that era of strict etiquette, there were many opinions about the correct and polite way of calling on the phone – what exactly should one say to draw the other person's attention to their instrument?

Graham Bell had suggested the greeting 'Ahoy!' which had traditionally been used for centuries by sailors as a greeting as well as to hail across from ship to ship. Bell himself refused to use any greeting other than "Ahoy" over the telephone throughout his life.

Bell's rival Thomas Edison would have nothing to do with "ahoy". Instead, he preferred to say "Hello", and insisted that anyone who wished to phone him should follow this example. Thus we can see here a faint echo of the "current wars" in which Edison and his rivals championed the use of Direct and Alternating current systems respectively.

On August 15, 1877 Edison wrote to T.B.A. David, president of the Central District and Printing Telegraph Company of Pittsburg, USA. David was just preparing to start a telephone service in Pittsburg, using equipment supplied by Edison's company. The historic letter reads:

"Friend David, I do not think we shall need a call bell, as **Hello**! can be heard 10 to 20 feet away. What you think? Edison - P.S. first cost of sender & receiver to manufacture is only $7.00." [11] This short message contains the first documented use of "Hello" with this particular spelling. Which raises the interesting question, how did Edison come up with this word?

"A Firm and Cheery Hulloa"

"Hello" was not used as a casual greeting until fairly recent times. According to the Oxford English Dictionary, the first use of "hello" was in 1827, less than 200 years ago. However, similar words like "hulloa", "halloo" and so on had long been used to catch someone's attention from a distance, to encourage a hunting dog to run faster, to summon a vehicle and so on.

By the mid-nineteenth century, "Hullo" was used to express surprise or to point out something remarkable and unexpected. This usage can be seen in works by Charles Dickens and other popular novelists. Even Sherlock Holmes exclaims, "Hullo, what's this?" when he finds a vital clue in the story "Silver Blaze". When Edison tested his phonograph recorder for the very first time, he spontaneously shouted "hullo" into the instrument.

When the first telephone directory was printed in New Haven, Connecticut, USA in 1878, it contained instructions for use along with guidelines on telephone etiquette. Users were encouraged to start their phone call with "a firm and cheery 'hulloa.'" By 1880 Edison's preferred hello had won out.[11]

Telephone in India

In early 1881, the Oriental Telephone Company Limited of England opened telephone exchanges at Calcutta (Kolkata), Bombay (Mumbai), Madras (Chennai) and Ahmedabad. On 28th January 1882, the first formal telephone service was established with a total of 93 subscribers.

From 1902 onwards, there was a gradual shift from cable telegraph to wireless telegraph, radio telegraph and ultimately radio telephone – a trend which was also seen in India over the decades.

On 25th November, 1960, India's first STD (Subscriber Trunk Dialing) call was made between Lucknow and Kanpur. Until that time, the telephone was a luxury and one had to book a trunk call and then wait, often endlessly, for an operator to arrange the connection. Later Indian telecom moved to digital, microwave, optical fiber, and satellite links.

The telecom revolution of India began to take off in the early 90s, as the DoT (Department of Telecommunication) ushered in a vision of accessible, affordable and modern communications for the common man. The first step was to install a large number of booths equipped with national and international calling facilities, providing employment to small entrepreneurs and physically challenged people while bringing the world to every neighborhood. From the mid to late 90s India used to talk through these STD and ISD (International Subscriber Dialing) equipped Public Call Offices (PCOs). The PCO booths opened up new vistas for millions of Indians for whom long distance communication options had so far been limited to tedious and expensive trunk calls, or else telegrams sent from post offices.

The industry saw a meteoric rise from almost zero to a thriving ecosystem within 3 to 4 years, with close to 1.5 million people employed in close to 1 million booths across India. DoT of India came out with standard regulations and also created a commission based revenue stream for the owners of the booths. The advent of these booths also gave birth to a parallel industry that produced industry standard billing system based on pulses.

Where outside of India, many countries offered calling card technologies, India persisted with these booths as the labour costs were low and these jobs were sought after. The introduction of these PCO booths had been a revolution and seen quick adoption in cities, towns and gradually in many villages as well. A large population of India used to completely depend on these facilities, but the PCO vanished quickly once the mobile revolution took off.

Figure 26: Haridwar, India 2006.
(Image Source: Flicker, Public Domain)

Shri Sam Pitroda a US-based NRI (Non-resident Indian) and a former Rockwell International executive is considered to be responsible for the telecommunication revolution in India and specifically, the iconic yellow-signed public call offices (PCO) that quickly brought cheap and easy domestic and international public telephones all over the country.

In 1984, Pitroda was invited to India from the US by the then Prime Minister Indira Gandhi. On his return, he started the Center for Development of Telematics C-DOT, an autonomous telecom R&D organization. In 1987, he became an advisor to Indira Gandhi's successor, Rajiv Gandhi and was responsible for shaping India's foreign and domestic telecommunications policies.

"This was in early 1970s. I was about 31–32 years old. I started a switching business with two friends in Chicago area. We built a business called Westcom Switching, and sold it to Rockwell International in 1979. It looked like so much money; I said what am I going to do with all this money?

I went to India just to visit. I had never been to Delhi. I went to Delhi, tried to call my wife from a 5 star hotel to Chicago and couldn't really make a good connection. I came back and I told my wife I found my mission in life.

I want to go fix India's telephones. Someone said "Hey you need to meet Mrs. Gandhi if you really want to do something on that scale". I didn't know how to meet Mrs. Gandhi, I had no connections, I didn't know anybody in Delhi. Finally through somebody, this man was a member of parliament from Gujarat, who knew my father in law, Mr. Baria, a very nice gentleman.

He organised a meeting and it was initially going to be a 10 minute

Figure 27: Sam Pitroda in 2014 (Image Source: Public Domain)

meeting and I decided to cancel it. I said look in 10 minutes you can't do anything. This is really serious, she should give me an hour and everybody said "Are you crazy? Why would Mrs. Gandhi give you an hour?" I said look we cannot do much in 10 minutes.

So I waited. Finally got a call after 8-9 months that she would give me an hour. I went back from Chicago with a presentation and all that. She had her entire cabinet there, everybody finance Minister Pranab Mukherjee, Venkatraman they were all there at that time."

-Sam Pitroda [19]

Private telecom companies began to enter the market in a big way from the late 90s, offering mobile telephone services that were cheaper than landlines. Gradually, telephone booths began using mobile phones for STD calls. Mobile phones were adopted by individual users, at first from the affluent set and then even by the middle class. Over time, mobile tariffs dropped to the point where they became the preferred option even for the lowest income groups. By around 2010, India became one of the world's biggest cell phone markets. National calling rates fell to the point where phone booths became an extinct species.

8

Wireless Communication

The future of radio is not in the radio. It is in the invisible waves

– Guglielmo Marconi

The study of electricity and magnetism accelerated quickly over the 18th and 19th centuries. Many great names, from Volta and Galvani to Franklin, Oersted, Ampere and Faraday, are connected with different milestones on this journey. By the mid 19th century, a large amount of experimental knowledge had been gathered. Based on these experimental facts, a handful of disconnected physical laws had been thought of and expressed mathematically. However, there seemed to be many missing links between various, apparently different, electrical and magnetic phenomena.

James Clark Maxwell had a vision of combining all the known aspects of electricity and magnetism into a single mathematical framework, and he succeeded in this during the 1860s. One outcome of this was that he could conclude that electricity and magnetism are linked together in such a way that their vibrations could travel long distances in the form of waves. Maxwell provided convincing evidence that light itself consists of electromagnetic waves, and he predicted that these waves could have a wide range of wavelengths besides light.

Heinrich Hertz was inspired by Maxwell's work, and he took up the challenge of demonstrating electromagnetic waves in the laboratory. He succeeded in 1888.

Maxwell and Hertz's work in turn kindled the interest of many physicists and other experimenters all over the world, such as William Crookes and Oliver Lodge.

The Indian scientist Sir Jagadish Chandra Bose, during his study at Cambridge, had witnessed the excitement around this new science. After returning to India, he pioneered many instruments to produce, transmit and receive electromagnetic waves with very short wavelengths. He carried out this work in Calcutta during 1895-1900. Bose was one of the first to realize that the shorter the wavelength, the more sharply a beam can be focused. Shorter wavelengths also give well-defined patterns when they are combined together in experiments. He chose very short wavelengths, from 2.5 centimeters to 5 millimeters, for his experiments and produced more definite and clear results compared to those achieved by his others at the time. He developed an innovative series of electromagnetic transmitters, detectors and antennas that could work at millimeter wavelengths.

Figure 28: Sir Jagadish Chandra Bose (Image: Public Domain)

Bose was successful in operating a remotely controlled switch over a distance of 25 meters, which in turn fired a pistol and rang a bell. He demonstrated this to the then Lt. Governor in India, Sir William Mackenzie, at a public event in Calcutta as early as 1895. This particular experiment by Bose was, perhaps, the first one of its kind. However, Bose later left the field of wireless research and turned to other scientific areas that interested him. [8]

Another pioneer was a young Irish-Italian named Guglielmo Marconi, who had a vision of using electromagnetic waves for long distance wireless signaling. After a series of short-range experiments, Marconi succeeded in spanning the Atlantic Ocean on December 12, 1901. He was successful in sending the letter 'S' in Morse code across the Atlantic from Poldhu in the UK to St. Johns in Newfoundland, Canada.

In recent years, historians have concluded that Marconi's landmark transatlantic success was based on a radio receiver that had been previously invented and published by Jagadish Chandra Bose. [20]

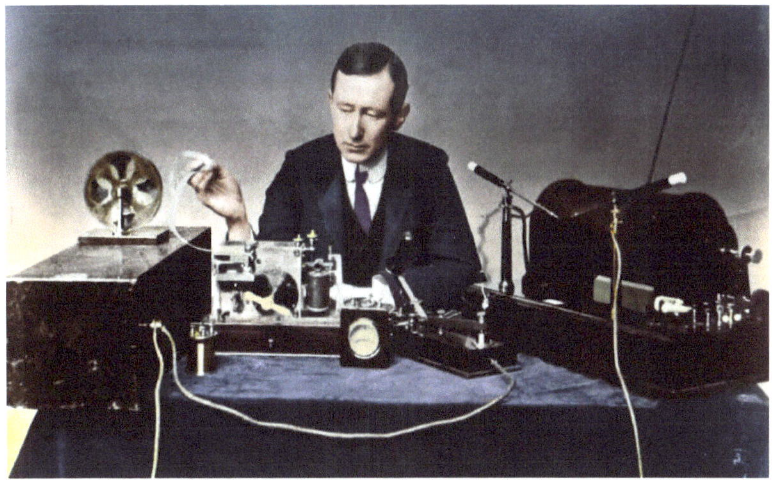

Figure 29: Guglielmo Marconi. (Image: Public Domain)

Marconi was not only an inventor but a shrewd and ruthless businessman. He fought many legal challenges and successfully defended his portfolio of key wireless patents from other claimants. One case against Nicola Tesla, dragged on for 43 years and was finally decided in Tesla's favour in the United States Supreme Court.

Wireless telegraphy was the first means of wireless communication and it continued up to World War I, when wireless-telephony was developed. It was adopted by military forces, civil aviation, merchant shipping, police, diplomatic services and others. This technology found a huge mass-market application in radio broadcasting as well.

The World Wars drove the development of smaller and smaller parts and equipment. The two-way communicators called "Handie Talkies" gained historical importance for their role in World War II. The "Walkie Talkie"—a bulky, 16-kg FM radio device with a range of 10 to 20 miles, was worn like a backpack and sometimes required two people to operate. These were to be the first steps in the evolution of portable personal communications that peaked in the modern mobile phone.

Figure 30: A portion of an early Motorola Radio ad, showing the handheld two-way communicator.
(Image Credit: Motorola)

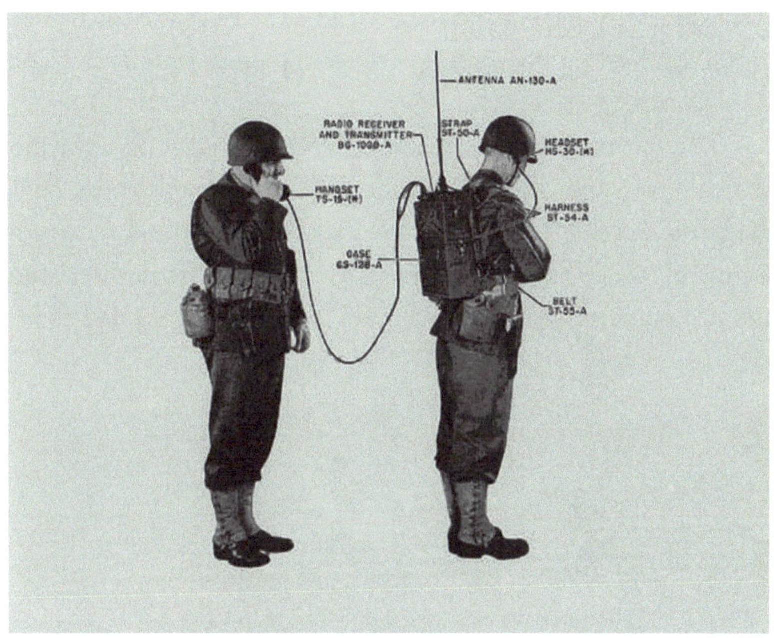

Figure 31: A diagram showing the proper use of a Walkie-Talkie (Image Credit: Motorola)

Wireless Telegraphy and the Titanic

RMS Titanic set sail on her first and last voyage in 1912. The liner boasted the most advanced features in marine architecture, safety, communications and passenger amenities. The Titanic's wireless room was equipped and staffed by Marconi's company. Chief telegraphist Jack Phillips and his assistant Harold Bride were on duty to handle operational messages as well as to send and receive personal radio telegrams for the passengers.

April 15, 1912 dawned with a failure of the radio equipment. Philips and Bride worked throughout the day to get the station running again, and succeeded by late evening. By that time, a backlog of passengers' messages had accumulated, which they began sending immediately to Marconi's shore station at Cape Race, Newfoundland.

The effort of repairing the set and then dealing with the delayed messages was slowly exhausting the operators, and there were still many more messages to be sent and received. Many of these were unimportant messages (including one about a game of poker), but the wealthy passengers had to be kept happy.

Figure 32: Titanic Sinking by Willy Stöwer, 1912
(Image Source: Public Domain)

Meanwhile, other ships had spotted dangerous icebergs in the area and were sending urgent radio warnings to the Titanic. The SS Mesaba informed Phillips of an ice field located directly in Titanic's path and waited for confirmation that Phillips had relayed the message to his captain. Philips did not reply to these warnings. The SS Californian also reported to the Titanic that the former was trapped in an ice field and had come to a stop.

Phillips replied impatiently, "Working Cape Race, keep out," and ignored this warning as well.

Phillips failed to report these warnings to the captain. At 11:40 p.m the Titanic struck an iceberg, flooding six of her sixteen watertight compartments. The liner began to sink, and Philips sent distress messages on the radio to nearby ships, including the RMS Carpathia. The Carpathia arrived a couple of hours later and could rescue 710 people out of the estimated 2200 passengers and crew. [12]

Edison and His "Spirit Phone"

In the late 19th century, side by side with keen enthusiasm for science and technology, there was an interest in what was then called the Spiritualism movement or "Psychical Research". This was the study of life after death, and the possibility of making contact with "departed souls". Many people hoped that these two areas could be combined - that is, scientific principles and techniques could be used to communicate across the veil of death itself. Famous names associated with this movement included William Crookes (physicist and pioneer of vacuum tubes), Oliver Lodge (physicist and early wireless pioneer), Sir Arthur Conan Doyle (creator of the fictional detective Sherlock Holmes) and many others.

Figure 33: (Image Credit: wwwfreepik. com/author/macrovector)

Nicola Tesla was also led to speculate on this possibility when he picked up some weird signals while experimenting with a radio receiver one night in 1901. He noted in his diary, "My first observations positively terrified me, as there was present in them something mysterious, not to say supernatural, and I was alone in my laboratory at night."

Figure 34: Portrait of Edison (Image by Abraham Archibald Anderson, Public Domain)

In 1918 he wrote in a more skeptical tone about such phenomena. "The sounds I am listening to every night at first appear to be human voices conversing back and forth in a language I cannot understand… I find it difficult to imagine that I am actually hearing real voices from people not of this planet. There must be a more simple explanation that has so far eluded me."

Today, it is well known that electrical storms and atmospheric disturbances can produce eerie howling, wailing or whistling sounds in a very low frequency receiver like Tesla's. In fact, these phenomena are actually referred to as whistlers, growlers, dawn chorus and so on. When such sounds are mixed together, it can be easily interpreted as many unearthly human voices talking, weeping or bemoaning some unhappy fate.

Edison was also influenced by such reports. In 1920, he told The American Magazine, "I have been at work for some time building an apparatus to see if it is possible for personalities which have left this Earth to communicate with us." This hypothetical device later came to be known as "Edison's Spirit Phone".

According to the authors of Edison vs. Tesla: The Battle Over Their Last Invention[13], Edison put a prototype of his spirit phone invention to the test in 1920. He invited both mediums and scientists to come over and observe a mysterious experiment. They saw a projector-like machine, set out on a workbench, that emitted a thin beam of light onto a photoelectric cell. The illuminated cell was meant to detect the presence of forces and objects moving through the beam— even those invisible to the naked eye. If a being from another world were to attend the gathering and pass through the light, a meter hooked up to the photoelectric cell would let them know, Edison explained. The witnesses waited patiently for hours, but nothing remarkable took place, and a disappointed Edison had to end the session.

Edison faced criticism for this aspect of his R&D work, and was even accused by some of practicing a deliberate hoax. However, extracts from his personal diary show that he himself was a sincere believer. He continued working on his so-called "spirit phone" throughout the 1920s.

Edison died in 1931 without producing results any more convincing than Tesla's earlier ones. But his determination was so strong that he planned an experiment that was to continue either after his own death, or after that of his engineer William Walter Dinwiddie. The idea was that whoever died first would try to make contact with the other. Dinwiddie passed away in 1920, about a decade before Edison. There is no record of the ultimate result of this particular experiment.

Figure 35: *Edison's burial place behind his home in West Orange, New Jersey. He has yet to communicate from there. (Image Credit: Bogdan Migulski/ WikimediaCommons)*

Later on, other people came forward with claims of ghostly messages that they had received from Edison after his death. A group of researchers claimed making contact with him during a séance in 1941. Edison's spirit, they reported, had finally cracked the spirit phone problem and provided details on how to build one. The group quickly constructed an instrument according to the ghostly blueprints, but they could achieve no more success than Edison's earthly attempts. [14]

9

Ham Radio or Amateur Radio

In ham radio, you don't just communicate; you connect

– Bill Smith

In the very early days of radio, any curious experimenter could build a radio transmitter and try to contact other radio enthusiasts using any frequency they liked. This became a popular hobby in Europe, USA and many other parts of the world. This community of amateur radio experimenters came up with many technical innovations that found their way into professional radio engineering practice. For example, it was amateur radio operators who first found that short wave radio could cover long distances, a fact which was later found to be due to reflections from certain layers of the Earth's atmosphere (later identified as the ionosphere).

Marconi himself can be considered the first "amateur" radio operator to get on the air — way back in 1901 with transatlantic communications. Other amateurs took to the airwaves in the early 1900s and the first "wireless" club was formed at Columbia University in 1908.

Figure 36: One Amateur radio station.
(Image Source: Wikipedia, Public Domain)

But this Wild West situation could not last long. An unregulated chaotic situation on the radio bands would interfere with critical communications and also pose a security risk. National laws and international agreements quickly came into force, and different frequency bands were regulated strictly.

The amateur radio community's usefulness as a technically skilled group of citizens, was recognized by reserving specific radio bands for their use. The hobby came to be known informally as "Ham" radio. Some consider the word ham to be a contraction of "amateur", while others cite the letters "HAM" as the call sign originally used by the Harvard Radio Club.

Since then, many notable milestones have been achieved by these hobbyists. More than 15 years before man was to walk on the moon, Ross Bateman, W4AO, and Bill Smith, W3GKP, bounced 2-meter signals off the moon from a station in Virginia in 1953 — and moon-bounce was born.

Amateur radio caught up quickly with the space age. The first of the OSCAR (Orbiting Satellite Carrying Amateur Radio) satellites were launched in the sixties, and hams could experiment with worldwide communications via satellite. In 1983, astronaut Owen Garriott, W5LFL, communicated with hundreds of hams while aboard the Space Shuttle high above the earth. It has become a tradition for manned space missions to hold special events where hams could interact with astronauts using their amateur stations.

Ham Radio in India

Amarendra Chandra Gooptu (2JK), licensed in 1921, was the first amateur radio operator in India. In the same year, Mukul Bose (2HQ) became the second ham operator. They were the first to make a two way amateur contact within India. By 1923, there were twenty British hams operating in India.

Figure 37: Former Prime Minister Rajeev Gandhi was seen operating Ham Radio in Vijayawada during 1990 May Cyclone (Image Credit: Academy of HAM radio)

During the Indian independence movement, Indian amateur operators turned their skills towards supporting the Independence movement, setting up clandestine pro-independence radio stations in the 1940s. Post-independence growth in the number of operators was rather slow until the then Prime Minister of India and amateur radio operator, Rajiv Gandhi (VU2RG), waived the import duty on wireless equipment in 1984. As of 2023, there are about 22,000 hams in India. The community has played a vital role during disasters and national emergencies such as earthquakes, tsunamis, cyclones, floods, and bomb blasts, by providing voluntary emergency communications in the affected areas.

The combination of fun, social, and friendship is the characteristic of ham radio. However, the mystique and charm of instant long range communications has decreased, now that mobile phones and internet access are commonly available. The hobby still has some limited appeal to electronics buffs, but it faces an uncertain future.

10

Cellular Phone

The cellphone has become the adult's transitional object, replacing the toddler's teddy bear with a mobile device

- Danielle Steel

In 1917, Finnish inventor Eric Tigerstedt filed a patent for a "pocket-size folding telephone with a very thin carbon microphone". Early predecessors of cellular phones included analog radio communications from ships and trains. The race to create truly portable telephone devices began after World War II, with developments taking place in many countries. The advances in mobile telephony have been traced in successive "generations", starting with the early zeroth-generation (0G) services, such as Bell System's Mobile Telephone Service and its successor, the Improved Mobile Telephone Service. These 0G systems were not cellular, supported few simultaneous calls, and were very expensive.

Figure 38: Image Courtesy: https://www.interviewgig.com

1G: Mobile Analog Voice

Cell phones began with 1G technology in the 1980s. 1G is the first generation of wireless cellular technology. 1G supports analog voice only. The maximum speed of 1G technology is 2.4 Kbps.

2G: Text Messaging Introduced

The 2G telephone technology introduced call and text encryption, along with data services such as SMS, picture messages, and MMS. The maximum speed of 2G is 64 Kbps.

3G: Mobile Data Introduced

The 3G is introduced 1998. It supported voice as well as broad band data such as video calling and mobile internet access. The maximum speed of 3G was around 2 Mbps for non-moving devices and 384 Kbps in moving vehicles.

4G: Mobile Data Focused

The fourth generation of networking, which was released in 2008, is 4G. It supports mobile web access like 3G does and also gaming services, HD mobile TV, video conferencing, 3D TV, and other features that demand high speeds. The max speed of a 4G network when the device is moving is 100 Mbps. The speed is 1 Gbps for low-mobility communication such as when the caller is stationary or walking.

5G: IOT Focused

5G is being deployed worldwide since 2019 promises significantly faster data rates, higher connection density, much lower latency, and energy savings, among other improvements. The anticipated theoretical speed of 5G connections is up to 20 Gbps per second.

History of Mobile Phone in India

1994: Private sector allowed in cellular and paging services.

1995:. W Bengal CM Jyoti Basu makes first call to Union Telecom minister Sukh Ram on July, 31.

2002: Reliance Infocomm starts CDMA mobile service

2004: More mobile phone than landlines.

2008: MNTL brings 3G to Delhi

2012: Kolkata first Indian city to get 4G.

2022: 5G services were officially inaugurated by PM Narendra Modi on October 1st, 2022

5G Roll-out in India

5G services were officially inaugurated by PM Narendra Modi on October 1st, 2022. Telecom industry titans — Reliance Industries Limited CMD Mukesh Ambani, Bharti Enterprises Chairman Sunil Bharti Mittal and Aditya Birla Group Chairman Kumar Mangalam Birla — shared the stage with the Prime Minister and committed to a speedy roll-out of "affordable" 5G services in the country. While Mr. Mittal announced the launch of Bharti

Airtel's 5G services in eight cities, including Delhi, Mumbai, Varanasi and Bengaluru, and plans to cover the entire country by March 2024, Mr. Ambani said Reliance Jio will begin the roll-out this Deepavali and committed to "deliver 5G to every town, every taluka, and every tehsil of our country" by December 2023. Mr. Birla added that Vodafone Idea will begin the journey to roll out 5G services soon.

The story behind the first mobile call in India

"It all started when sometime in mid-1994, Basu invited B K Modi, who was the chairman of the erstwhile Modi Telstra, and me in his office at the Writers' Building secretariat in Calcutta. We weren't expecting anything more than a courtesy meeting. Towards the end of the meeting, Basu, in his typical bhadralok manner, asserted that Calcutta should become India's first city to have a mobile network. The entrepreneurial zeal of Modi made him commit to an exact date: July 31, 1995, as my mind immediately moved towards 'project countdown'. Equipped with the determination to keep our word, we landed up in Australia to hold discussions with our joint venture partner, Telstra, to help us to find a suitable technology partner. The hunt for the technical expert who could roll out such a network brought us to Nokia, which had been a sleeping giant till then in Australia. Nokia had cutting-edge technology, but were initially reluctant. Perhaps it was the timeline. It took us a bit of convincing to get them going and it wasn't until Nokia agreed to accompany us that we hopped on the same flight back to India. And that's how we partnered with a sceptical Nokia to accomplish the impossible. Within nine months, the network was in place. We kept the promise by being ingenious, but played it by the book. There has been no looking back for Indian telecom since then."

- Umang Das, the then-CEO, Modi Telstra [29]

11

The Missed-Call Messaging

If you think about it, a missed call is often more than just an absence of voice; it's a presence of silence that speaks volumes.

– Paul Auster

Mobile talk time was very expensive in India in the early days. The thrifty Indian consumer soon realized that in many situations, it was not necessary to actually talk to the other person, but only to let them know that you have a message for them. In such cases, people would dial a number (often a landline) and then hang up immediately. The called party (using the caller ID feature that was also rolled out around that time) would then call back on the cheaper landline phone. The next level of missed-call ingenuity was to agree upon a specific pattern of missed calls in order to convey a specific message. Soon, public-facing corporate offices set up services such that customers could use a missed call to trigger an automated callback or text message.

Figure 39: A missed-call icon

Figure 40: Illustration of a ringing phone

Despite the fact that mobile talk time is much more affordable today, the missed call is still relevant in India. One example of this is that some companies acquire a large block of consecutive phone numbers that have no purpose other than to act as targets for missed calls. The private exchange hosting these numbers is set up to disconnect immediately, while sending the caller's number and the dialled number to a computer. As an example, this may act as a message to the computer to add a new subscription channel to the customer's DTH service, based on the last digits that they dialled.

This is very convenient for customers who may not own a smart phone or may not wish to install many apps on their devices. At the same time, the user experience is simpler than navigating a menu on an IVR (Interactive Voice Response) system, which would also incur an expense for the customer or for the provider.

12

Pager

The most important thing we learn at school is the fact that the most important thing is to keep in touch.

- Marvin Minsky

For a few years in the 1990s, pagers in India were the electronic status symbols of businessmen, doctors and executives. These devices could receive and display brief text messages, with top end models having two-way messaging capabilities. The sender would phone the service provider's call center and register the message request and the recipient's pager number. Then the message would be relayed by a wireless transmitter to the recipient's device. Pagers were generally worn on the belt or carried in the pocket.

Pagers operate as part of a paging system which includes one or more fixed transmitters (or in the case of response pagers and two-way pagers, one or more base stations), as well as a number of pagers carried by mobile users. These systems could be limited

Figure 41: *Alpha-Numeric Pagers. (Image Credit: Jtech)*

to the premises of a hospital, for example, with a single low power transmitter. Or, they could form a nationwide system with thousands of high-power base stations.

Pagers were developed in the 1950s and 1960s and became widely used by the 1980s. In the 21st century, the widespread availability of cellphones and smartphones has greatly diminished the pager industry.

These services were rolled out in India in 1995, Motorola being a major provider with nearly 80 percent of the market share. Mobilink, Pagelink, BPL, Usha Martin Telecom and EasyCall were the other players.

The business peaked in 1998 with the subscriber base reaching nearly 2 million. The subscriber base had dropped to 500,000 in 2002, roughly coinciding with the beginning of mobile telephony. The technology had become obsolete and practically disappeared by 2010.

13

Internet Communication

The Internet is becoming the town square for the global village of tomorrow.

– Bill Gates

As Stephen Hawking put it, "We are all now connected by the internet, like neurons in a giant brain." This observation is truer than ever today, with more than 4.95 billion users for whom the internet has become integral to daily life.

Figure 42: Image Source: http://blogs.lse.ac.uk/lsesadl/files/2014/01/social-media-cube670x335.jpg

Formally, the Internet is "a network of networks", a giant hierarchy of small networks linked into larger and larger networks to span the globe. Information is transmitted over the internet in the form of packets, which are routed automatically from sender to receiver based on the latter's internet address. The networking and routing systems don't really care about the actual content of the message, whether it is a web page or a digitized human voice or a video clip.

In this way the internet provides a readily accessible two way channel for almost any kind of information, and can take the place of dedicated specialized circuits like telephone networks. This has given rise to a host of internet based communications applications like email, Internet Telephony, WhatsApp, Skype, Google Meet, and Messenger. Once we have access to a broadband internet connection, we are a fraction of a second away from anyone with whom we may wish to communicate.

The Internet started in the 1960s as a way for government researchers to share information. Another catalyst in the formation of the Internet was the heating up of the Cold War. The Soviet Union's launch of the Sputnik satellite spurred the U.S. Defense Department to consider ways information could still be disseminated even after a nuclear attack. This eventually led to the formation of the ARPANET (Advanced Research Projects Agency Network), the network that ultimately evolved into what we now know as the Internet.

Internet History Timeline

1969: Arpanet

October 29, 1969, a message was sent from University of California in Southern California to a Stanford Research Institute computer console in Menlo Park, California. It read simply "Lo," though it was supposed to say "Login." The system crashed before completing the task. This was the world's first message sent via an interconnected computer network known as ARPANET. And thus the internet was conceived.

1971: Email

Email was first developed in 1971 by Ray Tomlinson, who also made the decision to use the "@" symbol to separate the user name from the computer name (which later on became the domain name).

1974: TCP/IP

Robert Kahn and Vinton Cerf developed TCP/IP protocol, a technology that links multiple networks together such that, if one network is brought down, the others do not collapse

1983: Arpanet Switch over to TCP/IP

January 1, 1983 was the deadline for Arpanet computers to switch over to the TCP/IP protocols. A few hundred computers were affected by the switch.

1984: Domain Name System (DNS)

The domain name system was created in 1984 along with the first Domain Name Servers (DNS). The domain name system was important in that it made addresses on the Internet more human-friendly compared to its numerical IP address counterparts. DNS servers allowed Internet users to type in an easy-to-remember domain name and then converted it to the IP address automatically.

1986: Protocol Wars

The Protocol wars began in 1986. European countries at that time were pursuing the Open Systems Interconnection (OSI), while the United States was using the Internet/Arpanet protocol, which eventually won out.

1988: IRC – Internet Relay Chat

Internet Relay Chat (IRC) was first deployed, paving the way for real-time chat and the instant messaging programs we use today.

1990: World Wide Web protocols

The code for the World Wide Web was written by Tim Berners-Lee, based on his proposal from the year before, along with the standards for HTML, HTTP, and URLs.

1990: The First Search Engine

Also in 1990, Alan Emtage, a college student in Montreal, created the first search engine for a school project. The search engine was known as the Archie Index.

1990s: The First Photo Shared on the Internet

Tim Berners-Lee shared the first photo on the Internet. It featured a group of singers known as Les Horribles Cernettes.

- 1991: First web page created
- 1991: First content-based search protocol, Gopher was launched
- 1991: MP3 becomes a standard
- 1993: Mosaic – first graphical web browser for the general public
- 1994: Netscape Navigator, Mosaic's first big competitor was released
- 1996: First web-based service, Hotmail was launched
- 1998: Google went live in 1998
- 2001: Wikipedia was launched
- 2003: Skype is released to the public, giving a user-friendly interface to Voice over IP calling.
- 2004: The Facebook open to college students
- 2005: YouTube – streaming video for the masses, Launched
- 2006: Twitter launched. The first Twitter message was "just setting up my twttr"
- 2007: The iPhone and the Mobile Web

Internet in India

The history of internet in India began with the launch of the Educational Research Network (ERNET) in 1986. The network was made available only to educational and research communities.

*Figure 43: Shammi Kapoor, one of the first Internet users in India
(Image Credit: BBC News 15 Jul, 2022)*

When Apple launched the eWorld service in June, 1994, they identified a few people across the world to hand out beta access accounts. In India these included Shammi Kapoor, Miheer Mafatlal, Vijay Mukhi, and K Pandyanand a few others. These users formed a group called the Internet UsersClub of India (IUCI) with Shammi Kapoor as the Chairman. On 28 July 1995, in an informal get together of the Bombay Computer Club (BCC), few IUCI members demonstrated the Internet to a gathering of about 40 people including B.K. Syngal, Chairman and Managing director of Videsh Sanchar Nigam Ltd (VSNL); Amitabh Kumar, director of technology at VSNL; Miheer Mafatlal and several other corporate honchos.

"I was fumbling with a rusted screw and people are cracking all sorts of jokes," he says. *"But finally, I connected, and placed the call from the modem. It creaked and croaked, if you know what I mean, and connected. A screen popped up, the home page of Apple eWorld."* K Pandyanand recalls. [9]

Few in the room knew what the Internet looked like. Fewer still knew about Apple's eWorld service, which had an email, a search box, some news and Talk City, a community chat room. Pandyan then asked people to step up and type any question and hit "enter". Some did. *"It showed answers as a Web page. And people went wow,"* he says.

However, Syngal and Kumar, gatekeepers of communication in India, were shocked rather than pleased at this demonstration. They were quite concerned that Pandyan could freely log onto a foreign computer network, apparently without the blessings and knowledge of VSNL, and that too by merely dialing a local number. At that time VSNL's own internet service had not yet been opened for public access. They noted the number Pandyan had dialed. After the demonstration Syngal went home, but Kumar went straight to the VSNL headquarters to investigate. It was only then that he found out that the number that gave them access was a VSNL one, connecting them to the British Telecom network. And a handful of users in India had been using that number to log on to eWorld.[9]

The first publicly available internet service in India was launched by Videsh Sanchar Nigam Limited (VSNL) on 15 August 1995. At the time, VSNL had a monopoly over international communications in the country and private enterprise was not permitted in the sector. The typical tariff in those days was around Rs. 40 per hour, with a minimum slab of 100 hours. This did not include the cost of the landline call itself.

In November, 1995, India's first cybercafé opened at the Leela Kempinski hotel in Mumbai through which hotel guests as well as outsiders could surf the net with a cup of coffee. It charged Rs. 800/- per hour. In December, 1995, a similar Cyber Café called Cyber Club was opened at the Maurya hotel in New Delhi.

Cyber cafes symbolized India's nascent internet revolution and introduced a generation to the World Wide Web. They were far less expensive than personal ownership of computer hardware or software, which were still seen as expensive luxuries by the middle class. By 2005, India had 200,000 cybercafes. Like the PCO and the pager, they became irrelevant and vanished when the smartphone became popular.

Figure 44: One cyber cafe of India, Twenty year ago
(Image courtesy: Hindustan Times, Jul 31, 2016)

In 2004, the government formulated its broadband policy which defined broadband as "an always-on Internet connection with download speed of 256 kbit/s or above. From 2005 onward, the growth of the broadband sector in the country accelerated. In 2014, India became the third largest internet users community with around 20 crore users.

Few Important milestones

- In 1996, Rediff.com was the first domain name registered in India
- In 1997, ICICI Bank launched the first Indian on-line banking site.
- In 1999, India ends VSNL's monopoly & Sify became India's first private ISP.
- In 2000, Webdunia, India's first Hindi portal was launched.
- In 2001 Indian railways launched online ticketing site.
- In 2003, Air Deccan becomes first airline to offer online ticketing in India

The India's Internet adventure appear to be highly positive. The emerging technologies like 5G, artificial intelligence, and machine learning will make Indian internet more customized and personalized.

Part-2

Satellite Communication & Beyond

14

The Birth of Modern Astronomy

The beginnings of astronomy were born out of humanity's curiosity about the heavens – our first step towards understanding the universe.

- Neil deGrasse Tyscon

In this chapter we look at the impact of the space age on the evolution of telecommunications. But first, we need to step backwards in time through a couple of centuries, because understanding natural satellites like the moon was a vital historical step on the way to launching the first man-made satellites, which then led to the era of satellite communications.

It was Galileo who first coined the term "satellite" to describe celestial bodies that revolve around the earth or around other planets. His

Figure 45: Galileo Galilei
(Image Source: Public domain)

observations using the telescope showed that the moon had geographic features like hills, valleys and craters and was thus a world of sorts rather than a perfect spherical object. He also recorded the moons of Jupiter for the first time. These astonishing discoveries helped to overturn the old idea that the earth was the centre of the universe.

Until Galileo's time, it was many still believed that the sun, moon and the stars revolve around the earth. Copernicus had already written in favour of the "heliocentric theory" placing the sun at the center with the planets orbiting around it, but his work gained little attention. Against much controversy, hostility and skepticism, Galileo championed the new picture of the solar system, which finally gained acceptance in the academic world. A new era of scientific astronomy began.

Figure 46: Sir Isaac Newton
(Image source: Public domain)

Johannes Kepler took the next step by stating that planets and their satellites follow elliptical orbits around their parent objects. He came up with a precise formula to calculate how these bodies would speed up and slow down as they moved along their elliptical orbits.

It was Isaac Newton who worked out the deeper reason for the elliptical orbits. He formulated the laws of gravitation and showed that the moon revolves around the earth for the same reason that a stone falls to the earth when dropped. One important logical conclusion was that if we could fire a cannonball skywards with enough velocity, then it could end up orbiting the earth just as the moon does – or it could even travel further out and reach other planets.

15

Space Flight in Science Fiction

Science fiction is the archaeology of the future.

- Clive Barker

The nineteenth century saw a mood of optimism and confidence in the ability of mankind to achieve the impossible by means of technology. The result was a rich vein of science fiction from the likes of Achille Eyraud (Voyage to Venus, 1863), Jules Verne (From Earth to Moon, 1865), Edward Everett Hale (The Brick Moon, 1869), and H.G. Wells (The First Men in the Moon, 1897). Edward Everett Hale in his book anticipated the ability to launch a satellite into so-called polar orbit. In his book, he described how an "artificial moon" could be deployed as a practical device for communications, earth observation, or navigation.

Many of these authors either had some level of scientific education, or had an interest in current scientific and technological developments at the peak of the industrial revolution. Their fiction was laced with a wealth of detail about the technical challenges that their protagonists faced, along with convincing technological concepts that enabled them to succeed in their adventures.

Of course, this interest in the outer universe was not focused only on humans traveling from the earth, but also included speculation about alien life forms arriving on our planet. H G Wells' War of the Worlds portrays a Martian spacefaring species that invades Earth with catastrophic consequences.

The science fiction waves of the late nineteenth and early twentieth century had a profound influence on the educated class of those times, and particularly on the children who read these works during their formative years.

Generations of youngsters grew up taking inspiration from these works, and some of them became the first pioneers of modern rocketry and spaceflight. Many an autobiography of a space scientist or engineer contains nostalgic recollections of a young mind ignited by these fantastic tales. For example, rocketry pioneer Robert Goddard was inspired by reading Verne and Wells as a teenager. German rocket engineer Wernher von Braun not only read science fiction in his youth but also wrote a novel about human flight to Mars.

16

The Modern History of Satellite Communication

Satellites have become the eyes and ears of humanity in space, revolutionizing the way we communicate, navigate and observe our planet.

– Stephen Hawking

By the twentieth century, technology was evolving very rapidly. The internal combustion engine combined with the work of the Wright Brothers made possible the dream of powered flight. Soon, a few visionaries began to look even further, considering ways for man to reach the moon and the planets. In Russia, Konstantin Tsiolkovsky began groundbreaking work on the theory and design of rockets that could carry people into outer space.

Robert Goddard was an early American rocketeer who successfully tested liquid-fueled rockets by 1926.

During World War II, the German Government assembled a team of scientists to develop rockets with explosive payloads. They built upon the concept of liquid fueled engines that Goddard had pioneered. A series of increasingly sophisticated weapons was developed, starting with buzz-bombs (which were rocket powered unmanned winged aircraft). This was followed by the V-1, and the V-2 rockets. The V-2 was the world's first large-scale liquid-propellant rocket vehicle, the first ballistic missile, and the first supersonic vehicle.

Towards the end of the war, the United States and the Soviet Union scrambled to capture German scientists and research facilities which would give them a head start in the postwar era. The space programs of both nations got a major boost because of the experts and technology that they acquired from Germany. A notable example was Wernher von Braun, a German scientist who was taken to the USA under this special recruitment program, also called "Operation Paperclip". Later he became a key figure in US rocketry and space programs – first in the Army and later in NASA.

Von Braun also helped develop the ATS series of satellites, one of which was "lent" to India as part of the Satellite Instructional Television Experiment (SITE) program in the seventies.

The potential use of man-made satellites for global communications was first studied by Arthur C. Clarke. He had worked during World War II in the British Radar Establishment. He came up with the idea that a satellite placed in a so-called "geostationary orbit" could act as a radio relay station covering about one third of the globe. Three such satellites would be sufficient to establish a global satellite-based relay network. In October 1945, he published his ideas and detailed technical calculations in the journal Wireless World.

Figure 47: Sir Arthur C Clarke (Image Credit: Arthur C Clarke Trust)

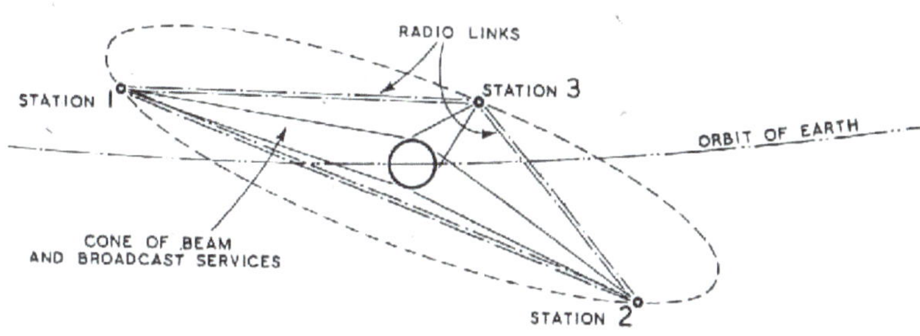

Figure 48: *Arthur C. Clarke's sketch of how just three satellites could cover the entire world.*
(Image Source: Public domain)

Figure 49: *Sir Arthur Clarke (Extreme right) with the SAC/ISRO installation team in front of his house in Sri Lanka*
(Image Credit: SAC)

At the time, most people (including Clarke himself) thought of this as an interesting idea of mainly academic interest. The possibility of actually launching satellites to orbit the earth was considered to be a futuristic dream which might perhaps come true only after many decades. He also thought that his "space stations" would have to be manned by maintenance crews, based on his experience with the unreliable radio components of that time. As a result, he did not apply for a patent on this concept.

However, events were to unfold at an unimaginable pace. Sputnik-1 orbited the earth just 12 years after Clarke's paper, and Yuri Gagarin made the first human space flight a few years later. Telstar, the first commercial communications relay satellite, was launched in 1962.

On October 7, 1957, the Space Age began with the launch of Sputnik-1 by the Soviet Union, the world's first artificial satellite. It was a silver sphere, the size of a beach ball, that orbited the Earth for three months. A radio transmitter on board sent pinging sounds down to Earth, which even amateur radio operators could hear.

Sputnik-1 caused much anxiety in the USA, where the Government and general public now feared that they would be overtaken and eclipsed by the Soviet Union in terms of scientific, technological and military capabilities. The Cold War and proliferation of nuclear weapons had added to the sense of urgency around the space race, since much of the technology was common between space vehicles and ballistic missiles. The USA accelerated their efforts. Four months after Sputnik-I, the United States launched Explorer-I on 1 February 1958. It operated for four months, transmitting scientific data back to earth.

Figure 50: A replica of Sputnik 1
(Image Source: Public Domain)

In December 1958, the US Signal Corps launched the world's first "broadcasting satellite." This satellite, known as SCORE, simply repeated a brief message from President Eisenhower: "Peace on Earth, Goodwill towards Men". The broadcast was on the air for several days, going silent when the batteries ran out.

The first communications satellite, Courier 1B, was launched by US in October 1960. It was designed to relay teleprinter traffic and had a capacity of 16 channels, proving the utility of orbiting satellites to relay communications signals.

In July 1962, sixty years after Marconi's first transatlantic message, NASA's Telstar-1 satellite achieved another transatlantic milestone - the first live television link across the Atlantic Ocean.

Figure 51: Model of a Telstar satellite
(Image Credit: Wikipedia. Public Domain)

The Relay satellite, built by RCA and launched on December, 1962, carried more experiments similar to Telstar. It continued in service through 1965.

The technical feasibility of satellite communications to support teletype, voice, and even television had been demonstrated by the end of 1962. The next big challenge was to show that a communications satellite could be successfully launched into geosynchronous orbit (sometimes called the Clarke orbit in honor of Arthur C. Clarke) and operated reliably from this great distance – almost a tenth of the way to the Moon.

Figure 52: Syncom-3
(Image source: Public domain)

The Hughes Aircraft Company tackled this challenge in 1963 with the Syncom satellites. Three satellites of this series were launched by NASA on Delta launch vehicles. The first launch was a failure, but the second, Syncom-2, was successfully launched in December, 1963. The Syncom-2 and Syncom-3 demonstrated that reliable communications to geosynchronous orbit with a return link to Earth was indeed technically and operationally viable. During the 1964 Olympics in Japan, television signals were transmitted from Japan to the USA via Syncom 3 and the signal was transmitted from the USA to Europe via the Relay-1 satellite. The era of live satellite broadcasting on a global scale had begun.

17

Satellite Communication in INDIA

We choose to go to the moon in this decade and do the other things, not because they are easy, but because they are hard

- John F. Kennedy

Inspired by Sputnik-1 and other developments that followed it, two young visionary scientists - Vikram Sarabhai and Homi Bhabha – decided that India, too, could benefit enormously from space technology. Both men had personal rapport with Prime Minister Jawaharlal Nehru, and they soon convinced him of the potential of space technology for national development.

The Indian government, in 1961, designated the

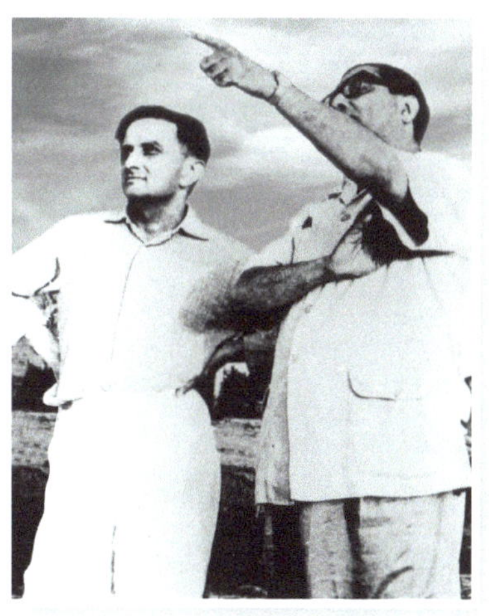

Figure 53: *Vikram Sarabhai and Homi J Bhabha*
(Image Credit: SAC Archives)

Department of Atomic Energy (DAE) as the organization that was to pursue the peaceful applications of space technology in India. They were to start with scientific studies of atmosphere, electrojet phenomena and astronomy. Dr. Homi Bhabha, the the secretary of DAE, in February 1962 set up the Indian National Committee for Space Research (INCOSPAR) with Prof. Prof. Vikram Sarabhai as its Chairman. INCOSPAR was to oversee all aspects of space research in the country, and its members were eminent scientists drawn from various disciplines. INCOSPAR's first R&D facilities were set up at Trivandrum (Rocketry and associated technology), and at Ahmedabad (space applications and associated instrumentation).

18

First Satellite Earth Station in India: ESCES

We are a nation of a billion people. We must find a way to harness the power of space technology to improve the quality of life for all our citizens

- Dr. Vikram Sarabhai

Dr. Sarabhai had succeeded in getting Government approval for the fledgling Indian space program, and now he needed to quickly set up the required infrastructure. The first step was a proposed Experimental Satellite Communication Earth Station (ESCES). This was set up on a hillock near Jodhpur village to the West of Ahmedabad. (This site ultimately grew into the campus of the Space Applications Centre). The story of how ESCES and SAC found their home is an interesting one, and was later narrated by Shri Jamnadas Patel, the first Chief City Town Planner of Gujarat State.

The Sarabhais owned some farmland near Jodhpur Village. Adjacent to this land was a hillock (or 'tekra' in Gujarati). Incidentally, the Sarabhai plot ended up as a snake park called Sundarvan.

Figure 54: A view of Ahmedabad Earth Station, ESCES)

In those days Dr. Sarabhai liked to invite friends to the farm for informal gatherings. On such an occasion in 1963, he confided to his friend Shri Jamnadas Patel that he was desperately looking for a piece of suitable land to start the INCOSPAR programme and to build up infrastructure like ESCES. Now, offers to host the program were coming in from many Indian states, but Dr. Sarabhai's fond wish was to see it taking root in his own birthplace – and if possible, close to his beloved farmland retreat. Unfortunately, his requests to the Gujarat Government for suitable land had not proved fruitful up to that point. So, during one of his farmhouse picnics, Dr. Sarabhai sought Shri Jamnadas's good offices to put in a word with the then Chief Minister Shri Balwantrai Mehta.

At 9:30 PM, that same night, Shri Patel telephoned the Chief Minister and requested him for an urgent appointment to discuss the matter. The CM did not sound very pleased, as he was just getting his bedtime massage after a very busy day. Could Mr. Patel come over next morning? Sorry, no Sir, insisted Jamnadasji – "It's very urgent, and needs your immediate attention!"

Finally, the CM consented to a brief meeting that night at 11:00 PM. The outcome of the meeting was that the CM agreed to review a more detailed plan the next day. Jamnadasji then rushed back to Dr. Sarabhai, and they spent the rest of the night drawing up a sketch of the proposed site. Next morning, Shri Patel presented the plan to the CM. A quick glance at the plan, and the CM jotted down his approval on margin, with an instruction that the land should be made available in 15 days.

A few days later a jubilant Dr. Sarabhai burst into Shri Jamnadas's home and, with a hearty hug, told him that the Gujarat Government had offered him the land at a token price of one rupee. In fact, they had been ready to provide him nearly twice the area indicated in the plan, encompassing all of the hillock, but Dr. Sarabhai accepted only the portion that he felt was actually needed.

The United Nations Development Programme (UNDP) provided assistance of $500,000 for setting up this Earth Station and nominated the International Telecommunication Union (ITU) as the executing agency for this project. A 14-meter dish antenna was set up with equipment from Nippon Electric Company, Japan. ESCES became operational in 1967. ESCES was used to train over 200 engineers, both from India and other developing countries in satellite communication and associated technologies. This historical earth station where the foundation of satellite communication in India was laid still stands firm in the campus of Space Applications Centre, Ahmadabad.

Figure 55: Prime Minister Smt. Indira Gandhi at ESCES in April,1968
(Image Courtesy: SAC Archives)

Figure 56: Dr. Vikram Sarabhai explaining to dignitaries at ESCES in April, 1968
(Image Courtesy: SAC Archives)

19

The First Agreement with NASA

Space technology is a tool for the development of our nation. We should utilize it for our rural and agricultural development.

- Dr. Vikram Sarabhai

In the 1960s, NASA planned an experiment to broadcast television programs directly to communities via satellite. The country which would receive these broadcasts would have to be large enough and also close to the equator for testing a direct-broadcast satellite. India, Brazil and China were shortlisted as potential regions for these broadcasts.

China was not formally recognized at the time by the U.S, and was therefore dropped. Brazil was ruled out as its population was concentrated in the cities, which would not get the most benefit from the widespread satellite footprint. Finally India was chosen as a tropical country with a widespread population. However, its strained relationship with the U.S. prevented the U.S. government from directly inviting its participation in the project. It was preferred that India might make the first request for assistance in the context of its own nascent space program.

India was already interested in the developmental and educational applications of space technology, and requested UNESCO to run a feasibility study for a suitable first project. Between 18 November 1967 and 8 December 1967, an expert UNESCO team worked in India. The result was a report proposing a pilot project in the use of satellite communication, stating that such a project would be feasible.

A study team of three engineers from India visited USA and France in June 1967, and confirmed that India could meet the technical requirements for the project. The Indian government then set up the National Satellite Communications Group (SATCOM) in 1968 to explore the possible applications of a geosynchronous communications satellite for India. This group consisted of representatives from various cabinet ministries, ISRO and All India Radio (AIR) And Doordarshan. They recommended that India should use the ATS-6 satellite – a second generation satellite developed by NASA – for an experiment in educational television.

NASA's director of international programs was Arnold Frutkin. He helped Dr. Vikram Sarabhai to interact with other NASA officials and to have more discussions on the proposed experimental broadcasts. Sarabhai saw this as a great opportunity for India to expand its space program and to train Indian scientists and engineers. In 1969, the Indian Department of Atomic Energy and NASA signed an agreement to cooperate in an experiment using a space satellite to bring instructional television programs to rural India.

Figure 57: Dr. Vikram A. Sarabhai and Dr. Thomas O. Paine, NASA Administrator signing the agreement.
(Image Courtesy: SAC Archives)

20

Formation of ISRO

We must be second to none in the application of advanced technology to the real problems of man and society

- Dr. Vikram Sarabhai

In 1969, space activities were shifted out of the Department of Energy and entrusted entirely to INCOSPAR, which was now scaled up and reorganized as the Indian Space Research Organisation (ISRO). Dr. Sarabhai took over as its first chairman. After the untimely demise of Prof. Sarabhai on 30th December 1971, Prof. M G K Menon became the chairman of ISRO in addition to his other responsibilities e.g., Secretary, Department of Electronics and the Director, TIFR. On June 1, 1972, the Govt. of India established a Space Commission. Prof. Satish Dhawan was appointed as Secretary, Dept. of Space and Chairman, Space Commission.

Figure 58: ISRO HQ Today
(Image source: ISRO.gov.in)

When Prime Minister Indira Gandhi picked Dr. Dhawan to head the space programme, he was the Director of the Indian Institute of Science in Bangalore, but was on a teaching sabbatical to his alma mater, California Institute of Technology. Dr. Dhawan accepted the offer but did not want to leave his teaching responsibilities. He stipulated that the new Department of Space would have to be headquartered in Bangalore. Both his conditions were accepted. Dr. Dhawan was allowed to continue as the Director of IISc. He drew his salary from IISc, accepting only **a token salary of one rupee from ISRO during his tenure there.**

21

Formation of Space Applications Centre

Space technology is not an end in itself; it is a means to an end, the end being the development of society

– Prof. Yash Pal

Once Bangalore became the headquarters of ISRO/DOS, the center of gravity of the space programme shifted away from Ahmedabad. Up to that point, the programme had taken shape mainly in Ahmedabad, where all the technical developments were still happening. The Physical Research Laboratory (PRL) founded by Dr. Sarabhai was in Ahmedabad. The Experimental Satellite Communication Earth Station (ESCES), was also established in the city in 1965, and it became the nucleus for work related to space communications. Therefore, when ISRO was set up, it had been generally assumed that Ahmedabad would be its headquarters. Instead, on Prof. Dhawan's insistence, Bangalore became the headquarters of DoS and ISRO.

Four independent units had been operating at Ahmedabad - Electronics Systems Division (ESD), Microwave Antenna Systems & Electronics Division (MASED), Remote Sensing and Meteorology Division (RSMD) and Electronics System Division (ESD). They were operated from make-shift sheds in the campuses of the local engineering college and a polytechnic. Two residential flats were also being rented from housing societies named Leena Apartments and Chitrakoot Apartments.

Figure 59: SAC Campus
(Image: SAC.gov.in)

Prof. Dhawan consolidated all these activities into the Space Applications Centre (SAC) - with a focus on application aspects of Space. He persuaded Prof. Yash Pal, a senior professor at the Tata institute of fundamental Research (TIFR) Bombay to lead this new center, whose mandate would be to realize Prof. Sarabhai's vision of space technology as a powerful tool for education and development. Prof. Yash Pal later recalled:

"Once there was a meeting in Ahmedabad and professor Dhawan came and held my arm and said: 'Yash, you keep on talking about doing this and doing that. Why don't you leave fundamental science

for five years and come to Ahmedabad? We are going to set up Space Applications Centre and somebody has to do it'. Though I had an offer to become chairman, UPSC, I thought what will I do at UPSC at 44? Then Dhawan pressed me and I came to ISRO. At that time there was an office at Sahajanand College. There was Chitnis and Karnik. There was wing commander Rao and others. All were busy planning for SITE." [4]

Prof. Yash Pal took charge as the first director of the newly set up Space Applications Centre. During his tenure he laid a strong foundation for the growth of space technology and applications, setting the pace for the center's contributions under many future directors.

SAC is a unique centre responsible for building satellite payloads designed as required for each mission. It also develops the ground infrastructure needed for various applications. SAC works hand in hand with user agencies to demonstrate the applications and deploy the complete solution. The technology areas under SAC's mandate include Satellite Communications, Navigation and Remote Sensing. SAC is staffed by multi-disciplinary teams comprising space technologists, application developers and social scientists.

22

SITE Programme

The world is changing rapidly, and television will be one of the instruments to shape that change in India.

- Indira Gandhi

The First Space Application Programme in India.

In 1969, the Indian Department of Atomic Energy and NASA signed an agreement to cooperate in an experiment using a communications relay satellite to bring instructional television programs to rural India. Under the agreement, the Satellite Instructional Television Experiment (SITE) was launched in 1975. The USA made the Application Technology Satellite (ATS-6) available to India for this purpose. Direct Satellite Reception Systems were installed in 2400 villages for community viewing of television programs, spread out over six states of India – Rajasthan, Bihar, Orissa, Madhya Pradesh, Andhra Pradesh and Karnataka. The TV programmes were broadcast for four hours every day with educational / instructional content on family planning, agriculture, national integration, teacher training, occupational skills, health and hygiene etc.

Figure 60: *Village children watching TV programme with Prof. Yash Pal (Image Courtesy: SAC)*

SITE was a collaborative programme lead by ISRO with the active participation of Doordarshan (at that time AIR), the national TV broadcasting organisation, the Ministry of Education, etc. The project was supported by various international agencies such as the UNDP, UNESCO, UNICEF and ITU.

ISRO established a TV Studio at Ahmedabad as the main studio for SITE programme broadcasts. Special education programmes were to be made at another studio in Bombay.. Space Applications Centre (SAC), Ahmedabad was the nerve centre of the entire SITE program as all the major facilities like Earth Station, Studio and SITE program management office etc. were located there. Prof. Yash Pal took five years' deputation from TIFR, and went to Ahmedabad to build up the Space Applications Centre almost from scratch.

"I had never worked this hard; I did not know what happened. Firstly, the Satellite Instructional Television Experiment (SITE) was very much delayed. In fact, I remember some people came to see me as soon

as I was there and told me that people are saying that those who have joined the SITE team better get off before the ship sinks! And that kind of talk was going around. The Americans who were to provide the satellite for the experiment were telling us that we were not getting ready, and it was true that we were not getting ready." – Prof. Yash Pal [4]

However, Prof. Yash Pal had a remarkable capacity to face any challenge and could inspire his team to succeed even under the most difficult situations. Work on SITE soon took off in full swing and the project was successfully completed.

The SITE transmissions had a very significant impact in the Indian villages. Throughout the one year duration of the program, thousands of villagers gathered around TV sets installed at community centers. Studies were conducted on the social impact of the experiment and on viewership trends.

Figure 61: *The UHF antenna used for direct reception of TV programme from the satellite*
(Image Courtesy: SAC)

The SITE story: Delhi Earth Station

One back-up earth station was needed in Delhi for the SITE project. The original plan was to import the equipment, but the procurement time did not match with the project schedule. Prof. Yash Pal and his team decided to build the station in India.

Prof. Yash Pal and his team surveyed the Delhi ridge area in search of an open and elevated location for the earth station. They found a place in the forests adjoining Malcha Marg, but the Indian Air force objected to the proposal. They were worried that the station's signals would interfere with their communication or navigation systems. Finally, the then Prime Minister Indira Gandhi called a meeting with the Air Force officials. Prof. Yash Pal came over from Ahmedabad armed with technical data to show that the fears of interference with aircraft were groundless. At the meeting, he was able to prove convincingly that the fears were groundless, and Mrs. Gandhi decided then and there that the earth station would be set up at the proposed site. However, even Mrs. Gandhi was compelled to make one concession to the Air Force – the facility would be allowed to function only for a year, during the period of the experiment. It was to be dismantled at the end of the SITE program. As planned, the Delhi station played its part during SITE, and it continues to exist as an ISRO center to this day.

Some of the equipment – the high-power amplifiers, low noise amplifiers and so on, which were designed and built at SAC at that time, are still working at this earth station. SAC engineers built several earth stations in later years, including the Master Control Facility at Hasan in Karnataka for the control of the INSATs and other satellites.

Figure 62: Delhi Earth Station
(Image Courtesy: SAC)

Figure 63: The Prime Minister of India, Smt. Indira Gandhi, with Prof. Yash Pal and Prof. E.V. Chitnis during her visit to the Delhi Earth station set up by SAC for SITE.
(Image Courtesy: SAC)

The SITE Story: Low Noise Amplifiers

Many more challenges were to crop up before the indigenous earth stations could be completed. One example worth mentioning is the development of the low noise amplifier, which was to be designed by SAC engineers. The Low Noise Amplifier is installed at the focus of the dish antenna to boost the extremely weak satellite signal. Today, LNAs are off-the shelf items that you might have noticed on your satellite TV dish. In the mid-1970s there were no companies manufacturing them anywhere in the world, and it was a formidable task to design them in India. The plan was that NASA would provide the technical specifications, and SAC would design the units according to those specifications. SAC would provide six prototypes, after which the Electronics Corporation of India (ECIL) would produce the hundreds of LNAs that would be installed in the dish antenna at each community center.

Figure 64: An artist's view: Hot & Cold test of SITE-LNA, by Dr. Abhijit Chatterjee

One important specification was that the LNAs were to perform correctly over a large temperature range, since they would be mounted outdoors facing summer, winter and rain. At that time SAC was a new centre and was not well equipped with standardized environmental testing facilities. They resorted to makeshift test procedures such as placing the unit in a domestic refrigerator for low temperature testing, or outdoors in sunlight for high temperature testing. After completing the tests at SAC, the units were to be sent to NASA for final technical approval for use in the program.

Some eight units of LNAs were duly sent to NASA. Within a few days, NASA reported that the units were not at all satisfactory because they were very sensitive to the fluctuations in temperature. NASA recommended that ISRO should send their engineers to Goddard Space Flight Centre in the USA to work with NASA engineers to improve the LNAs and test them thoroughly.

Prof. Yash Pal was naturally worried after getting such feedback from NASA, but he did not give up. He held a meeting with the LNA team and went over all aspects of the design. It soon emerged that the problem was related to certain cost-cutting decisions in selecting electronic components for the LNA. Due to this, the unit was unable to compensate for fluctuations in temperature. Prof. Yash Pal invited other experts from India for an urgent two-day brainstorming session on how to improve the LNA. Within a week some eight new units were built with the modified design and again sent for testing to NASA. Within a few days a telex arrived with good news – the new units had passed the tests; they were fantastic!

"This I consider to be a very important moment in building self-confidence," said Prof. Yash Pal. He believed if the SAC engineers had been denied the chance, it would have been difficult to develop the atmosphere of creativity that came to pervade the SAC." – Prof. Yash Pal [4]

The SITE story: The Pij TV Transmitter

During the planning phase of SITE, it was felt that, in addition to the satellite broadcasts, a low-power transmitter could also be put up for local TV broadcasts. So a low-power transmitter was designed at SAC. A transmitter tower was needed, which would have to be linked to the Ahmedabad studio via microwave. But the question was: where to put the transmitter? At that time there was no TV centre in Ahmedabad. If the transmitter were to be set up very near to Ahmedabad, there would be a strong demand for regular TV services and not the rural programmes developed for SITE. Prof. Yash Pal called a meeting of his team. He knew that the transmitter would have a range of about 50 kilometers. Taking a Gujarat map and a compass, he drew a circle with a radius of 50 km around Ahmedabad. He pointed out a village called Pij just outside this circle, some 60 km south of the city in the Kheda district of Gujarat. Dr. Pal penciled a little circle around the village. Then, tapping the map, he declared, "That is where we will put our transmitter!" [4]

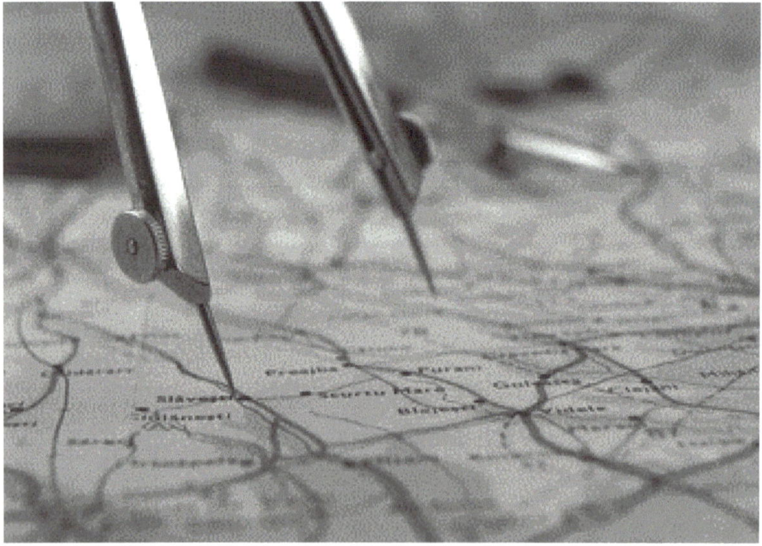

Figure 65

Another bureaucratic hurdle cropped up almost immediately, just as it had happened with the Delhi site. This time, the objection came from the Posts and Telegraphs department. They were worried because the proposed TV transmitting antenna was to be mounted on the same tower as the microwave link receiver connecting to Ahmedabad. They feared that the TV transmitter would interfere with the microwave equipment. Once again, the SAC team had to demonstrate that these concerns were groundless. The transmitter and microwave link were duly put up at Pij, making it the first Indian village to enjoy a television broadcast service. Programmes were broadcast in Gujarati, along with national programmes from Delhi via satellite.

After the completion of SITE, the program transmission continued as the Kheda Communications Project (KCP). For ten years the people of Kheda district in Gujarat keenly watched independent television telecast by the Space Applications Centre (SAC), beamed from a transmitter in Pij village. The programmes were relayed from local stations to the 651 community TV sets installed by milk cooperatives and district panchayats. They focused on raising awareness of health, farming, finance, and other development issues relevant to the villagers' lives. KCP was undoubtedly a success

Figure 66: The Pij TV Transmitter (Image Courtesy: DECU)

as a development catalyst and was awarded the UNESCO-IPDC (International Programme for the Development of Communication) award for rural communication efficiency in 1984. After the commissioning of the high-powered Doordarshan transmitter in Ahmedabad, on July 25, 1985, the state Government shut down the Pij transmitter.

The SITE story: The Amul Connection

Figure 67(a)

Figure 67(b)

It had been estimated that some 500-600 TV sets would be needed to equip the participating Kheda villages, but the main

SITE / KCP budget did not cover the cost of procuring them. Prof. Yash Pal invited Dr. Verghese Kurien, Chairman of Amul Dairy, to visit SAC. He proposed that SAC could work closely with Amul on producing programmes on animal husbandry, dairy farming and other topics for farmers. Dr. Kurien responded with enthusiasm and agreed that this was an excellent idea. Prof. Pal then gently introduced the question: perhaps Amul could bear the cost of the TV sets? Dr. Kurien agreed immediately, and Amul did fund the installation of the TV sets in hundreds of villages. Veterinarians from Amul were involved in making numerous programmes and in turn, they learned from SITE studio crews how to make TV programmes.

The SITE story: SAC TV Studio

There was not much experience in India at that time in building studios for TV production. The SAC team took on the task of setting up their own studio at Ahmedabad to produce the Pij programs, and arranged for a microwave link from the studio to the Pij transmitter. Up to that point, the only TV studios in India had been located at New Delhi and were run under the Ministry of Information and Broadcasting (I & B).

Yet again, a new bureaucratic hurdle had to be negotiated at this stage by Prof. Yash Pal and his people. Until then, the I & B Ministry was the only authority that had been mandated to run broadcasting studios and transmitters in India. They were adamant that this privilege could not be extended to any other arm of the Government. For his part, Prof. Pal was keen to build a creative, autonomous production team as part of SITE and KCP, which would continue such work in the long term while based at Ahmedabad. His vision was that the technical and production teams should work in concert with the sociological and educational researchers in a holistic manner under the overarching banner of space applications for social development, through the coming decades.

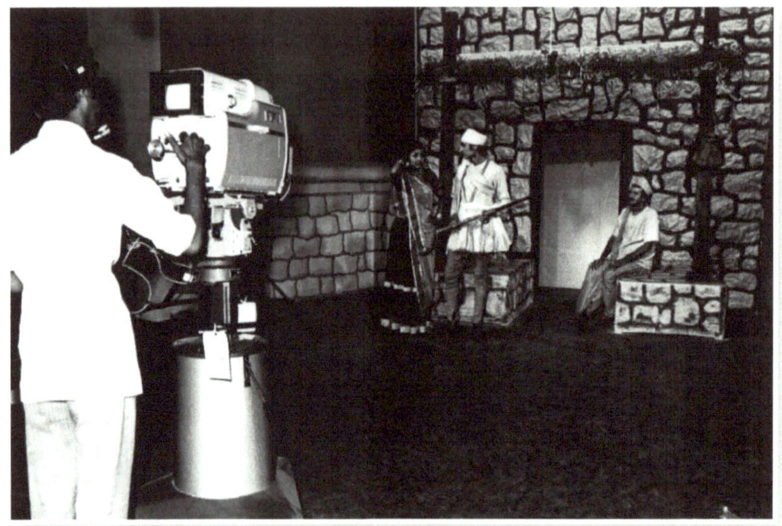

Figure 68: The SAC TV Studio
(Image Courtesy: DECU)

The SAC team held several fruitless meetings with the mandarins from the I & B ministry, until Prof. Pal finally threatened to quit the SITE project altogether. The stalemate could only be resolved when Prof. Satish Dhawan himself intervened and worked out a compromise of sorts with the Ministry officials. All India Radio would post a Station Director at Ahmedabad, and the broadcasts would be deemed to be nominally under his direction. At the same time the SAC teams were, in practice, given the creative freedom that Prof. Pal wanted for them.

Figure 69: Inauguration of SITE-Prof. Satish Dhawan explaining
Hon'ble Minister V. C. Shukla
(Image Courtesy: SAC)

Figure 70: Inauguration of SITE-Hon'ble Minister Shri. V.C. Shukla,
Prof. Satish Dhawan & Shri. Pramod Kale
(Image Courtesy: SAC)

The end of SITE Programme

As per the original agreement, the SITE program ended in July, 1976. NASA shifted its ATS satellite away from India, despite demands from Indian villagers, journalists and others such as Arthur C. Clarke to continue the experiment.

Similar experiments were conducted in the Appalachian region, Rocky Mountains, Alaska, Canada, China and Latin America in the mid-seventies and early eighties. These experiments demonstrated that satellite TV could play a very important role in providing education.

23

STEP Project

STEP laid the foundation for a robust communication framework that would be crucial during natural disasters and emergencies, proving that technology can save lives.

- V. S. Hegde

SITE was followed by a project called Satellite Telecommunications Experiments Project (STEP), using the Franco-German satellite SYMPHONIE during 1977-79. Whereas SITE had been a satellite broadcasting experiment, STEP was a project to demonstrate telecommunications links using a geosynchronous satellite.

The Symphonie satellites were launched by the American Thor Delta 2914 satellite-launch vehicles. Their missions were subject to a restrictive agreement imposed by the US State Department: any commercial use of Symphonie was forbidden. Due to this constraint, the satellites were used for experiments all over the world that could be broadly classified as scientific/technical experiments and humanitarian / cultural /educational experiments. In fact, these satellites may ultimately have had a bigger impact on the future than they would have had, if they had merely been deployed for some specific narrow commercial application.

Transportable Remote Area Communication Terminal (TRACT)

Figure 71: TRACT
(Image Courtesy: SAC)

A Transportable Remote Area Communication Terminal (TRACT) was indigenously designed by the Space Applications Centre and formed a part of the STEP ground equipment. TRACT was primarily a mobile earth station mounted on a truck with an antenna assembly. STEP started with a live TV transmission using TRACT from Amreli in Gujarat.

The Indo-Pak Cricket Match

The "TV Origination Experiment" from TRACT was carried out during the Indo-Pakistan Cricket Test matches in 1978. This transmission from Lahore, Pakistan enabled the Indian viewers to see live cricket matches played in a foreign country, for the first time. [2]

Figure 72: live cricket matches played in a foreign country

Emergency Communication Terminal (ECT)

Another important experiment in STEP was to demonstrate the use of a small jeep transportable / air lift-able terminal to provide speedy and reliable communication via satellite for supporting relief work during emergencies. A mobile satellite terminal called the Emergency Communications Terminal (ECT) was built and various experiments were conducted under STEP to demonstrate the ECT's potential role in emergency communications. Many simulated emergency operations were conducted including jeep transportation, air-lifting by helicopter and commercial aircraft, antenna installation and operation in remote areas.

Figure 73: Emergency Communications Terminal
(Image Credit: SAC)

During the 1977 cyclone in Andhra Pradesh, the ECT was put to actual use in the relief operations.[2]

Figure 74: TRACT and Outdoor Broadcasting (OB)
Van surrounded by curious onlookers at Amreli
(Image Courtesy: SAC)

24

Aryabhata: India's First Satellite

The success of Aryabhata was not just a triumph of technology, but a testament to our resilience and determination to forge our own path in space.

- Dr. K. Kasturirangan

From 1969 onward ISRO was developing a 40 kg satellite compatible with India's first satellite launch vehicle (SLV-3), which was similar to the Scout launch vehicle of the United States.

In 1971, India's then Prime Minister Indira Gandhi received a message from the country's ambassador in Moscow, saying that the Soviet Academy of Sciences was ready to assist India in launching its first satellite. Since China had already launched their first satellite in 1970, the Russian proposal became very attractive to India at that time.

India sent a team with Prof. U.R. Rao, who was in charge of satellite activities in India, to the USSR to discuss and finalize the details of this joint satellite programme. The Soviets initially suggested that since satellite technology is complex, India should go step-by-step with their programme. Their advice was first to make only a satellite payload and put it on a Soviet satellite. However, Prof. U.R. Rao insisted that India would start with designing a complete satellite. But for his firm stance, India would have lost the lead in satellite technology by at least a decade. [6]

Although China was a latecomer to the space club, being fifth after USSR, USA, France and Japan, they did break previous records for satellite weight. The first Chinese satellite weight 173 kgs, or more than twice the weight of Sputnik.

The Soviet Union suggested that India should design a satellite larger than the Chinese one, rather than the original 40 kg version. ISRO scaled up the satellite to 360 kg satellite, making it much larger than the 173 kg Chinese spacecraft. [6]

As happened often in those days, the high level decisions and approvals came through suddenly as a result of behind-the-scenes negotiations. The formal signing of the Indo-Soviet agreement was scheduled at fairly short notice, and Prime Minister Indira Gandhi requested ISRO to provide a cost estimate for the satellite within 48 hours.

"We worked through the night to arrive at a budget estimate of Rs. 50 to 60 lakh, which appeared ridiculously low. Prof. Rao's highly intuitive mind made him suggest that we multiply all our estimates by a factor of 6 and the total budget request then came to a respectable Rs. 3 crore. Prof. M.G.K. Menon, who was leading the space programme temporarily at that time, submitted this request to

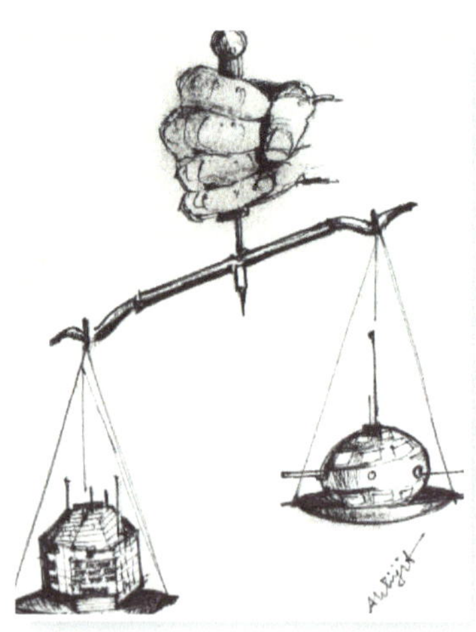

Figure 75: *Aryabhata vs. Dong Fang Hong 1 An artist's view by Dr. Abhijit Chatterjee*

Prime Minister Gandhi. Thankfully, the Aryabhata project was successful and actually required spending was also close to Rs. 3 crores".– K Kasturirangan.[6]

Until that point, the satellite had not even been named. ISRO had suggested three names for the satellite – Aryabhata, Maitri and Jawahar. Aryabhata was the first preference, and Mrs. Gandhi selected the name Aryabhata.

Figure 76: *Dis-Assembled mode test of Aryabhata (Image Source: ISRO.gov.in)*

Aryabhata was the first unmanned Earth satellite built by India. It was assembled at Peenya, near Bangalore, and launched from the Soviet Union on a Russian-made rocket in 1975. Aryabhata had instruments to perform scientific studies of the Earth's ionosphere, to measure neutrons and gamma rays from the Sun, and perform investigations in X-ray astronomy.

The scientific instruments had to be switched off during the fifth day in orbit because of a failure in the satellite's electrical power system. Useful information, nevertheless, was collected during the five days of operation. The Aryabhata's launch was successful at a time when leading space powers had little faith in India's chances to produce an indigenous satellite.

Figure 77: Aryabhata was helicopter-borne for testing the telemetry system in Sriharikota Range (SHAR).
(Image Courtsey: ISRO)

A commemorative postage stamp was issued by the Post & Telegraph department of India to mark the Launch of the First Indian Satellite, Aryabhata.

"Shankar Dayal Sharma who was in charge of our postal services (later he became the President of India) used to tell me, 'You made me spend double the money for printing the commemorative stamp on satellite.' This was because, though the launch date was fixed as 19 April 1975, I had told him that there could be a delay of one day. So he printed stamps both for 19 and 20 April 1975. But the launch did take place on 19 April itself and the stamp bearing 20 April 1975 had to be destroyed immediately." -U R Rao [21]

Figure 78: Image Courtesy: https://istampgallery.com

25

India's Own Launch Vehicle: SLV-3

You have to dream before your dreams can come true. We had to overcome many hurdles, but we were fueled by our dreams and commitment to the nation.

- Dr. A.P.J. Abdul Kalam

In the early 1970s, ISRO started the Satellite Launch Vehicle (SLV) project to develop the technology needed to launch satellites from within India. SLV was intended to reach a height of 400 km and carry a payload of 40 kg. It was a four-stage rocket with all solid-propellant motors. The first experimental flight of SLV-3, in August 1979, was a failure. The second flight was successfully launched on July 18, 1980 from Sriharikota Range (SHAR), when Rohini satellite, RS-1, was placed in orbit, thereby making India the sixth member of the exclusive club of space-faring nations. Dr. A P J Abdul Kalam was heading the project at that time.

"The year was 1979. I was the project director. My mission was to put the satellite in the orbit. Thousands of people worked nearly 10 years. I have reached Sriharikota and it is in the launch pad. The countdown

was going on…T minus 4 minutes, T mi nus 3 minutes, T minus 2 minutes, T minus 1 minute, T minus 40 seconds. And the computer put it on hold… don't launch it. I am the mission director, I have to take a decision", Kalam says in the video as he recounts the incident. [3]

However, Kalam says that the experts advised him to go ahead with the launch as they were confident about their calculations. Kalam decided to bypass the computer and launched the rocket.

"I bypassed the computer and launched the system. There are four stages before the satellite is launched. The first stage went off well, and in the second stage, it got mad. It went into a spin. Instead of putting the satellite in orbit, it put it into the Bay of Bengal", says Dr. Kalam.

However, Kalam said the decision was his and he was the one who took the call to overrule the computer warning. "First time I faced failure… And how to manage failure? Success I can manage, but how to manage failure?" Kalam says. ISRO chief Dr. Satish Dhawan held a press conference along with me.

Figure 79: *SLV-3 Liftoff*

"Dear friends, we have failed today. I want to support my technologists, my scientists, my staff, so that next year they succeed", said Dr. Dhawan.

He took the whole blame on himself despite criticisms. He took all the blame and assured them that next year we would succeed because his team was a very good one.

Next year, on July 18, 1980, the same team successfully launched Rohini RS-1 into orbit. Then Dr. Dhawan asked me to conduct the press conference that day.

"I learned a very important lesson that day. When failure occurred, the leader of the organization owned that failure. When success came, he gave it to his team". [3]

Balloon Based Communication

For the second flight, scheduled on 18th July 1980, Dr. Dhawan decided to allow Doordarshan to telecast the launch, but DD was not yet equipped with the technology for a live telecast.

Figure 80: Dr. APJ Abdul Kalam With Prof. Satish Dhawan & PM Mrs. Indira Gandhi in 1980.
(Image source: Prasar Bharati)

Engineers from SAC had come up with a unique solution. They had tethered a huge balloon with a transponder halfway between SHAR and Madras. Gummidipoondi was the chosen location for the balloon which floated at a height of about 1 km above ground. A long strong cable secured it firmly to the ground. The distance as the crow flies from SHAR was about 80 km. Experts from TIFR's balloon facility at Hyderabad had been roped in for this project. Yash Pal and Dr. Aravamudan were to provide the live commentary.

"Yash Pal, Gita and I were rehearsing our questions the previous day when we heard some bad news. The blimp had flown off! This kind of balloon was usually used in events like carnivals which took place in areas where there was a benign breeze that created just a gentle oscillation. No one had accounted for the strong winds which roared through Gummidipoondi and wrenched the balloon off its tether. Nothing could be done about it as the launch was scheduled for the next day, at dawn. We decided to record the commentary and rush it to Madras by road" - Dr. Aravamudan[7]

Stubborn Remote-Controlled Cable

There were two sets of umbilical cords attached to the Satellite Launch Vehicle (SLV-3). One set came off automatically at the launch while the heavier set was detached remotely. A few minutes before the take-off, the sequence had to be put on 'hold' because the cable refused to come off. A technician named Bapiah came to the rescue.

The launch vehicle was armed and hence very unsafe for anyone to approach. The technician volunteered to climb the launch tower and remove the cable. The tower was around 60 ft high, which was about the same height as the rocket. We had no other option but to let him try, with the safety officials turning a blind eye just for a minute. Bapiah quickly climbed the tower and gave the cable a hefty kick – and it mercifully came off! The rest, of course, is history.

"The flight was a maiden success — a milestone in ISRO's history. And so on 18 July 1980, almost seventeen years after the first foreign Nike-Apache sounding rocket was launched from TERLS, a made-in-India rocket launched from Indian soil injected an Indian-made satellite into a 300 km by 900 km orbit. It was an ecstatic moment. Kalam was hoisted on the shoulders by his colleagues. In Trivandrum we were all welcomed as heroes when we stepped off the plane. My little sons were thrilled. In their school the SLV had been dubbed the Sea Loving Vehicle. And now their father's organisation had been vindicated! The successful SLV-3 flight was a real morale booster to ISRO. India had become the sixth member of the exclusive club of space-faring nations. Decades have passed by since then and this club has not increased in strength!" - Dr. Aravamudan [7]

26

The First Geostationary Satellite of India: APPLE

The success of APPLE demonstrated India's potential to design and build its own satellite systems, an essential component of our communication infrastructure.

– Prof. U.R. Rao

The European Space Agency (ESA) had announced an opportunity for launching a satellite on its third developmental flight of their launch vehicle, Ariane-1. The conditions were that the satellite had to carry a European payload METEOSAT on its top and had to be delivered within three years. If the satellite was not ready in that time, a dummy satellite would have to be built and launched.

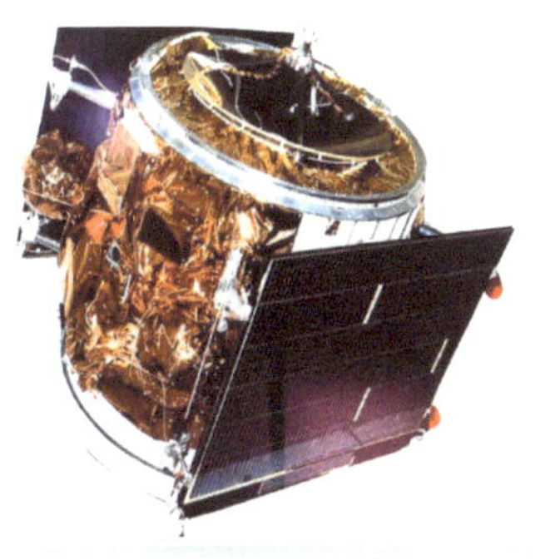

Figure 81: *APPLE*
(Image: ISRO.gov.in)

ISRO lost no time in submitting their proposal, which was chosen by ESA from among seventy-two competing proposals in 1976 and approved by Government of India in May 1977. The time schedule was thirty months with a budget of Rs. 21 crores. This was definitely a bold step with a very tight schedule. Thus, the first communication satellite, APPLE, was born.

APPLE was configured with two C-Band transponders. The Space Applications Centre (SAC), Ahmedabad was assigned the task of design and development of the payload. The payload was designed to be fully compatible with the ground segment already available through the earlier Satellite Telecommunication Experiment Project (STEP).

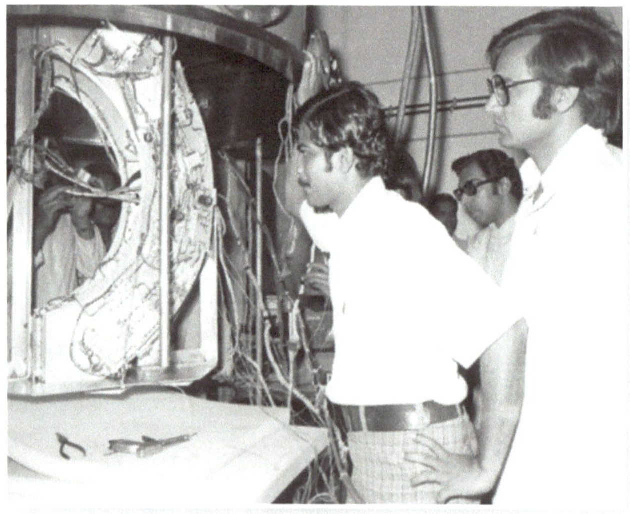

Figure 82: SAC Scientists & Engineers at work on the APPLE payload
(Image Courtesy: SAC)

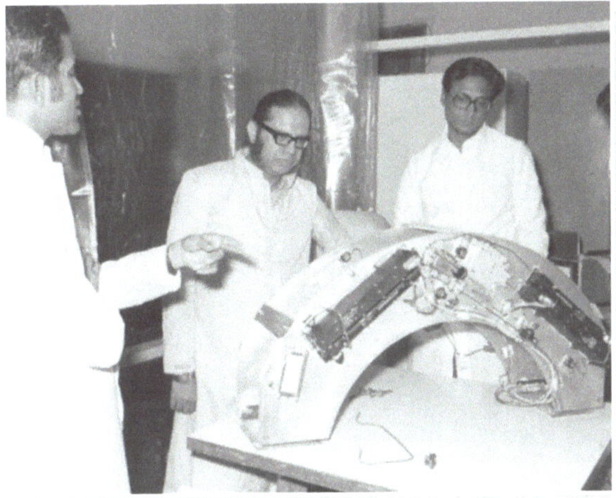

Figure 83: SAC Scientists & Engineers at work on the APPLE payload
(Image Courtesy: SAC)

Figure 84: Apple satellite being transported on a bullock cart.
(Image Source: ISRO.gov.in)

How Ancient Technology Saved the Day

When the satellite was ready to be transported to the launch pad at Kourou, French Guiana, some last minute glitches were noticed in the Telemetry, Tracking and Control (TT&C) links which are crucial for maintaining communication with satellites in space. It was decided that the antenna would have to be tested thoroughly to make sure that this link would function perfectly.

Today, ISRO is well equipped with antenna testing facilities located within "anechoic" (echo-free) chambers, but four decades ago such facilities did not exist in India. The team considered the option of transporting the satellite to Toulouse in France for final tests, but the expense and delay was found to be unacceptable.

As it turned out, the tests were completed in a few hours at a cost of ₹150. A local farmer was paid that amount to hire out his bullock cart to the Indian Space Research Organization. APPLE was placed on the cart and taken to an open field where tests could be run without echoes or interference. The TT&C link problem was quickly understood and solved, and the results were documented for technical approval to go ahead with the launch. [27]

APPLE in Orbit

The Ariane LO3 launch on 19 June 1981 from French Guyana was a perfect one and APPLE was released into a nominal Geosynchronous Transfer Orbit (GTO). The SHAR centre took control of the satellite immediately after launch. Of the two solar panels, one deployed correctly but the other did not. When all the efforts to deploy the second panel failed, it was decided to go ahead with only one panel. The C-Band transponder was switched on and for the first time, Ahmedabad Earth Station received signals from an Indian-built Communication Satellite.

Live Dance Program Broadcast via APPLE

The successful functioning of the APPLE transponder was tested using a live dance programme signal sent to and received from the satellite. APPLE transponder also transmitted Rabindranath Tagore's dance drama Chitrangada[27]

Figure 85: Live Dance Program

Video Conference with the Prime Minister

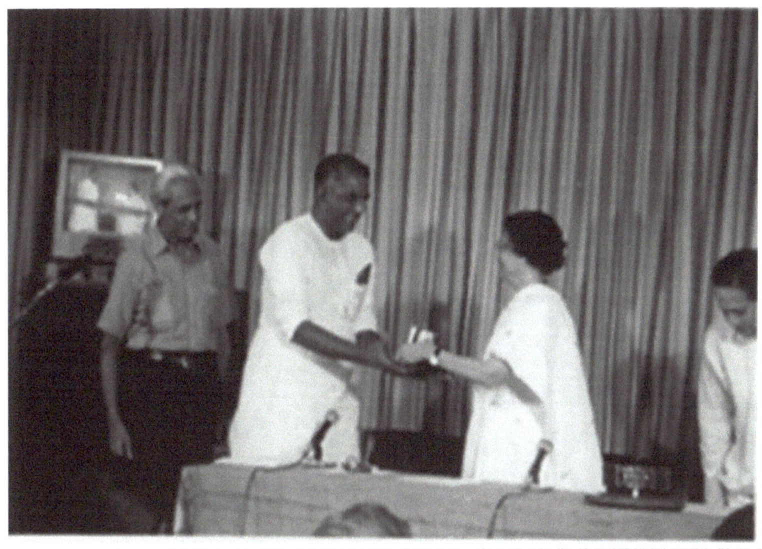

Figure 86: "I dedicate APPLE and our services, once more, to the people of India". - Prime Minister, Smt. Indira Gandhi.
(Image Courtesy: SAC)

APPLE was formally dedicated to the nation on 13 August 1981 by then Prime Minister Indira Gandhi. The APPLE development team members, assembled at the

Vikram Hall at Space Applications Centre, were introduced to the Prime Minister by Satish Dhawan, the then Chairman ISRO, who was at New Delhi Earth station, through two-way videoconferencing – one half of the screen showing the Delhi Earth station scene and the other half showing the Vikram Hall. In today's age of internet meetings we take this kind of event for granted, but it was a noteworthy technological step in the first days of APPLE.

The Prime minister symbolically handed over a model of APPLE to the Minister for Communications and said that APPLE marked the 'Dawn of India's satellite communication era'. The Prime Minister's address to the nation on 15 August from the Red Fort was also carried live to the country by APPLE.[27]

The Friend in Need

APPLE rose to the occasion when road and communication links were shut down due to a cyclone that hit Amreli district. The satellite was used to link Gandhinagar Secretariat and Amreli District Collector's office, with emergency communication terminals at both locations and ESCES acting as a hub station. More than 50,000 messages (from the birth of a child to bereavement to postponement of betrothal) passed through APPLE during the next 10 days until conventional links were restored. The Chief Minister of Gujarat visited SAC to express gratitude on behalf of the citizens and government of Gujarat. He said, *"When we gave our land at Jodhpur Tekra for ESCES we never knew that it is going to be extremely useful in our times of distress. You are our friend in need."* [27]

27

Indian National Satellite System-1 (INSAT-1) Series

Technology is the driving force of our development. The INSAT system has opened new avenues for communication and education, empowering millions of people across India.

- Dr. A.P.J. Abdul Kalam

INSAT-1 Series

The SITE and STEP experiments were the stepping stones towards the conceptualization of the Indian National Satellite System (INSAT). In the late sixties through early seventies, ISRO undertook several studies in collaboration with experts from several international organizations. These included Massachusetts Institute of Technologies (MIT) - Lincoln Laboratory (LL), NASA, and General Electric. The study team submitted a report in March 1971, including a tentative configuration for the INSAT spacecraft.

The INSAT system was conceived as an operational multipurpose and multiuser satellite system. The INSAT-1 satellite series was designed by ISRO scientists and engineers but was built by an American company, Ford Aerospace. Even though experience to build geosynchronous satellites had been gained through APPLE, the decision to have the INSAT-1 series built by foreign companies was driven by the need to establish quickly an operational system so that satellite communication became an accepted technology by the user agencies in the country than wait until the ISRO attained the maturity to build operational satellites. The decision was also influenced by the fact that occupation of geo arc over India and frequency spectrum was based on 'first come first served' basis, hence early occupation of geo slots through procured satellites was vital.

Figure 87: INSAT-1B Integration
(Image: Public Domain)

Ford was chosen from among three bidders, the other parties being Hughes Spacecraft and a European consortium. Only Ford offered to build a three-axis stabilized platform, necessary for meteorological cameras to work on a continuous basis. INSAT-1 satellites built by Ford Aerospace were launched from abroad through procured U.S. and French launches.

INSAT-1 Series satellites carried two S Band high power TV broadcast and twelve 12 C Band telecommunications national coverage transponders, in addition to meteorological services.

INSAT-1A was launched by the Delta vehicle from the U.S. on April 10, 1982. Following a series of failures, the satellite was abandoned in September 1982, less than 6 months into a seven-year mission.

INSAT-1B was launched on 30 August 1983. Full operational capability was achieved in October 1983. It continued to operate till 1990. It could simultaneously relay 4375 two-way phone calls. Around 36,000 earth images were returned. Eleven of its 12 C-band transponder and its two S-band transponders provided direct nationwide TV & communications to thousands of remote villages, along with a detailed weather and disaster-warning service. Around 35,000 indigenously built 3 to 3.6 meter diameter, receive-only terminals were in place to supply rural communities with social and educational programs.

By the end of the decade, almost 90% of the population and 80% of the Indian landmass of the country was covered by the Television through INSAT-1B.

The INSAT-1C satellite was launched on 21 July 1988 from Kourou, French Guyana. Half of the 12 C-band transponders and its two S-band transponders were lost when a power system failure knocked out one of the two power buses, but the meteorological earth images and its data collection systems were both fully operational. Earth lock was lost on 22 November 1989 and the satellite was abandoned.

The INSAT-1D, the last spacecraft of the INSAT-1 series, was planned for 29 June 1989. Unfortunately, 10 days before that, during launch preparation, a Launchpad hoist cable broke and a crane hook fell on the satellite and damaged its C-band reflector. The fully insured satellite was repaired by Ford Aerospace at a reported cost of $10 million. But that mishap was followed by solar panel damage costing $150,000 during the 1989 San Francisco earthquake. The satellite was finally launched on June 12, 1990.

The development of ground infrastructure had begun in step with INSAT-1A, continuing to grow rapidly after the launch of INSAT-1B. One or the other Doordarshan receive/transmit system was being inaugurated every day and the communication services were fast expanding. VHRR pictures were shown along with news bulletins on the weather. While SITE was an experiment for the unprivileged in certain select regions, INSAT-1B was bringing satellite communications to the public all over India. Important services like e-banking, National Stock Exchange and e-governance started to utilise satellite-based communication effectively. INSAT-1 brought new inspiration to society as a whole. While Aryabhata was an inspiration in terms of technological pursuit in India, INSAT-1 brought about a change in the mindset towards digital connectivity. It was now clear that the time had come for a fully Indian designed and built series of satellites to sustain and expand the valuable services that INSAT-1 program had introduced.

28

Indigenous Operational Satellite: INSAT-2 Series

The INSAT system has redefined communication in India, bridging gaps and connecting the remote with the mainstream.

- Dr. K. Radhakrishnan

INSAT-2 Series

ISRO engineers had been closely associated with the configuration and operation of the INSAT-1 series, and had also developed the APPLE satellite. They were now ready for the challenge of a fully indigenous operational multi-purpose communication and broadcasting satellite series – the INSAT-2 series of satellites.

Two INSAT-2 "test spacecraft" were approved by the Government to be built during 1980-1990, in order to ramp up the capabilities of ISRO to the new level of complexity prior to launching the actual operational series. However by the time INSAT-2A was realized, the confidence level was sufficiently high that the "2" series was deemed to be an operational one. INSAT-2B carried more transponders and featured the new extended C band frequencies for enhanced communication capabilities.

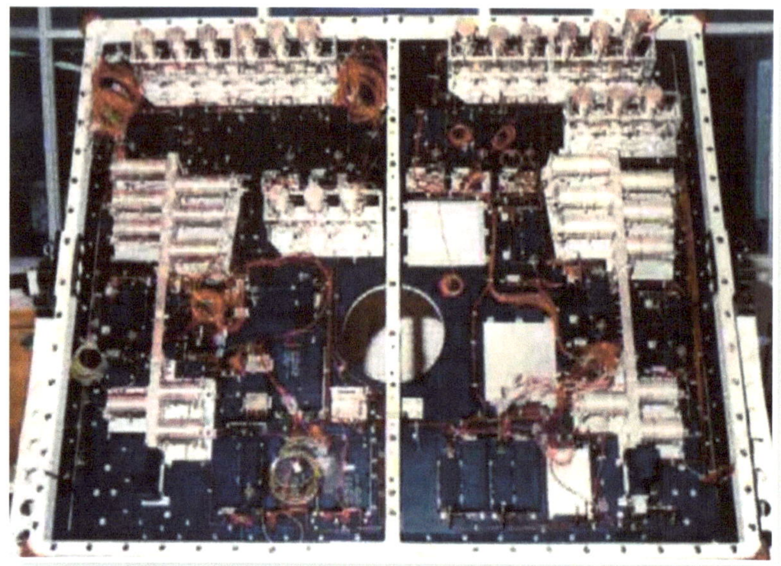

Figure 88: INSAT-2 ETM South panel
(Image Courtesy: SAC)

Before building INSAT-2A, an "Engineering and Thermal Model" (ETM) of satellite was built and tested. The communication payload design, fabrication and testing was done at SAC, Ahmedabad. Starting from the mid-eighties to late eighties, scientists and engineers at SAC were busy acquiring and establishing their advanced capabilities in terms of design, fabrication, integration and testing of complex payload hardware. The ETM model of payload was complete by 1988.

Prime Minister Rajiv Gandhi talks to President Reagan

A critical component, a Microprocessor Integrated Circuit, was needed for use in INSAT-2A's Attitude Control Electronics. At that time, the US Government considered this item to be so sensitive and strategically important that they did not permit its export even to their allies. ISRO had been trying for two years to procure this microprocessor.

Only after Prime Minister Rajiv Gandhi took it up directly with President Reagan, were the devices cleared for export to India. Even then, the US manufacturer's senior representative had to deliver them personally to ISAC, Bangalore after traveling with the package chained to his wrist.[28]

INSAT-2A was launched and commissioned in August 1992 with INSAT-2B following a year later to augment the capacity and also as a backup. The designation of "Test-Satellite" (TS) was dropped after the launch and these satellites provided very useful services to the nation as operational satellites. INSAT-2A and INSAT -2B served their intended life and the satellite communication in India entered into a new phase of exponential expansion.

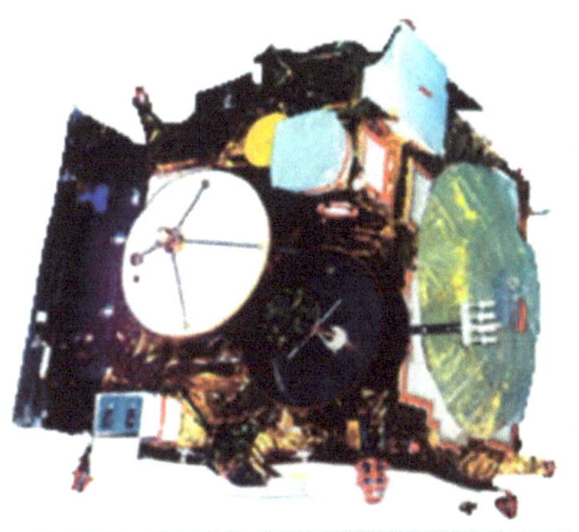

Figure 89: INSAT-2B Spacecraft
(Image courtesy: ISRO)

INSAT-2A and 2B in the early 90s became workhorses to expand telecommunication services, supporting V-SATs in extended C-band, TV through S-band broadcast and other telecommunication services in C band. Demand for satellite based services was growing far beyond anticipated levels. Full exploitation of the VHRR data in the numerical weather model had to wait till 2008, when a new era of IMD modernisation began. Nevertheless, Search and Rescue and Data Relay payloads made very good contributions.

Figure 90: Prime Minister Atal Bihari Vajpayee with the fully integrated INSAT-2E. (Image Courtesy: ISRO)

29

ISRO Today

The only way to reach the stars is to dream, and ISRO has taken us closer to those stars with every successful mission.

– Rakesh Sharma (Indian Astronaut)

ISRO has come a long way from the days of SITE, STEP and APPLE. 43 Communications satellites have been launched so far. Today the Indian National Satellite (INSAT) system is one of the largest domestic communication satellite systems in the Asia-Pacific region with nine operational communication satellites placed in Geo-stationary orbit. It initiated a major revolution in India's communications sector and became a key element of national infrastructure. The whole gamut of satellites includes Ku Band High Throughput Satellite, S Band Multi-beam Multimedia Satellite, C Band and Ku band Capacity enhancement Satellite and Multi-Frequency, Multi-coverage and Multi-function satellite for Strategic Users. The INSAT system with more than 200 transponders supports applications like telecommunications, television broadcasting, satellite newsgathering, societal applications, weather forecasting, disaster warning, Search and Rescue operations etc.

TABLE-1: List of communication satellites launched by ISRO

S.No.	Satellite	Launch Date	Launch Mass Kg	Launch Vehicle
1	APPLE	19-Jun-1981	670	Ariane -1(V-3)
2	INSAT-1A	10-Apr-1982	1152	Delta
3	INSAT-1B	30-Aug-1983	1152	Shuttle [PAM-D]
4	INSAT-1C	22-Jul-1988	1190	Ariane-3
5	INSAT-1D	12-Jun-1990	1152	Delta 4925
6	INSAT-2A	10-Jul-1992	1906	Ariane-44L H10
7	INSAT-2B	23-Jul-1993	1906	Ariane-44L H10+
8	INSAT-2C	7-Dec-1995	2106	Ariane-44L H10-3
9	INSAT-2D	4-Jun-1997	2079	Ariane-44L H10-3
10	INSAT-2E	3-Apr-1999	2550	Ariane-42P H10-3
11	INSAT-3B	22-Mar-2000	2070	Ariane-5G
12	GSAT-1	18-Apr-2001	1530	GSLV-D1
13	INSAT-3C	24-Jan-2002	2650	Ariane5-V147
14	INSAT-3A	10-Apr-2003	2950	Ariane5-V160
15	GSAT-2	8-May-2003	1800	GSLV-D2
16	INSAT-3E	28-Sep-2003	2775	Ariane5-V162
17	EDUSAT	20-Sep-2004	1951	GSLV-F01
18	INSAT-4A	22-Dec-2005	3081	Ariane5-V169
19	INSAT-4C	10-Jul-2006	2168	GSLV-F02
20	INSAT-4B	12-Mar-2007	3025	Ariane5
21	INSAT-4CR	2-Sep-2007	2130	GSLV-F04
22	GSAT-4	15-Apr-2010	2220	GSLV-D3
23	GSAT-5P	25-Dec-2010	2310	GSLV-F06
24	GSAT-8	21-May-2011	3093	Ariane-5 VA-202
25	GSAT-12	15-Jul-2011	1410	PSLV-C17

S.No.	Satellite	Launch Date	Launch Mass Kg	Launch Vehicle
26	GSAT-10	29-Sep-2012	3400	Ariane-5 VA-209
27	GSAT-7	30-Aug-2013	2650	Ariane-5 VA-215
28	GSAT-14	5-Jan-2014	1982	GSLV-D5
29	GSAT-16	7-Dec-2014	3182	Ariane-5 VA-221
30	GSAT-6	27-Aug-2015	2117	GSLV-D6
31	GSAT-15	11-Nov-2015	3164	Ariane-5 VA-227
32	GSAT-18	6-Oct-2016	3404	Ariane-5 VA-231
33	GSAT-9	5-May-2017	2230	GSLV-F09
34	GSAT-19	5-Jun-2017	3136	GSLV Mk III-D1
35	GSAT-17	29-Jun-2017	3477	Ariane-5 VA-238
36	GSAT-6A	29-Mar-2018	2117	GSLV-F08
37	GSAT-29	14-Nov-2018	3423	GSLV Mk III-D2
38	GSAT-11	5-Dec-2018	5854	Ariane-5 VA-246
39	GSAT-7A	19-Dec-2018	2250	GSLV-F11
40	GSAT-31	6-Feb-2019	2536	Ariane-5 VA-247
41	GSAT-30	17-Jan-2020	3357	Ariane-5 VA-251
42	CMS-01 (GSAT-12R)	17-Dec-2020	1425	PSLV-C50
43	CMS-02 (GSAT-24)	23-Jun-2022	4181	Ariane-5 VA-257
44	CMS-03 (GSAT-20)	19-Nov-2024	4700	Falcon 9 Block 5

Figure 91: ISRO in 1963: The photograph of the rocket nose cone on a bicycle taken by Henri Cartier-Bresson On the right: engineer, CR Sathya. His assistant, Velappan Nair, is taking care of the nose cone.
(Image Courtesy: ISRO)

Figure 92: *ISRO in 2022: ISRO's LVM3 to make commercial foray with launch of 36 OneWeb satellites.*
(Image Courtesy: ISRO)

30

Formation of Indian National Space Promotion and Authorisation Centre (IN-SPACe)

Space is not just a program; it is a part of our aspiration. By enabling the private sector, we are opening new frontiers for our youth and encouraging innovation that can change the world.

– Narendra Modi (Prime Minister of India)

Over the past two decades, private enterprises such as Virgin Galactic, SpaceX, Blue Origin and Arianespace have revolutionized the global space sector by reducing costs and turnaround time. By contrast, private enterprises in India have been limited to being merely suppliers to the government's space programme.

In 2020, the Government of India took the historic decision to encourage of private enterprises across all phases of space activities. The government also formed the Indian National Space Promotion and Authorisation Centre (IN-SPACe) with its headquarters in Ahmedabad.

Figure 93: PM Shri Narendra Modi with Shri Pawan Goenka, Chairman IN-SPACe during the inauguration of HQ of IN-SPACe at Bopal, Ahmedabad. (Image Courtesy: IN-SPACe)

IN-SPACE will operate within the Department of Space helping private players to become independent actors rather than being merely subcontractors or suppliers. The new Space Policy will allow private entities to own, develop and launch rockets and even to set up their own launch facilities. They will also be able to own and operate satellites in orbit. So far, imaging satellites are only owned by ISRO and Defense. ISRO will be mostly tasked with the R&D on cutting edge technologies as well as scientific and interplanetary missions. Currently, India constitutes 2-3% of the global space economy and is expected that with this reform this will increase to more than 10% by 2030.

31

Future Trends in Satellite Communication

I think it's possible for ordinary people to choose to be extraordinary. We're going to make space travel as common as air travel. I think it's important for humanity to be a multi-planetary species.

- Elon Musk (CEO of SpaceX)

Future communications will bring together not only people but also things, data, applications, smart cities, and transportation in a smart, integrated, and seamless networked society. In this context, it is envisaged the seamless integration of satellite and terrestrial networks and 5G is the first step towards that. It is also foreseen that four key technology elements of will dominate satellite communication in very near future, v.i.z., High Throughput Satellite (HTS), LEO Satellite constellation, Software-Defined Satellite & Satellite Based Secure Quantum Encrypted Communication.

High Throughput Satellite (HTS)

High Throughput Satellites has transformed the satellite communications industry in recent years. They are a new generation of spacecraft, capable of delivering vast throughput when compared to conventional satellite systems. These satellites are optimized for high data rate applications, using multiple spot beams with extensive frequency reuse, thus achieving greater capacity than that of conventional wide-beam satellites used for broadcast applications. This calls for a multi-beam satellite with spot beams covering the desired region, using high frequency bands such as Ku-band and Ka-band.

Many HTS systems have already been placed into orbit since the 1990s. They are deployed mainly in the Ku & Ka-bands. Launches will continue for the foreseeable future, especially in the larger and somewhat less densely occupied Ka-band.

ISRO already launched three HTS, GSAT19 (2017), GSAT-29 (2018) & GSAT-11(2018) with an aim to provide equal digital speed in rural and urban areas and bolster the overall internet speed in India.

Hughes Communications India (HCI) has launched India's first commercial high-throughput satellite (HTS) broadband service. Hughes will provide the HTS broadband using the Indian Space Research Organisation's (ISRO's) GSAT-11 and GSAT-29 satellites. With the new HTS capabilities powered by ISRO satellites, ISRO is confident that HCI will continue to deliver excellent quality satellite broadband services and further enhance the connectivity experience that accelerates India's digital transformation.

This new broadband service will address connectivity gaps, improve network performance, and support the high bandwidth requirements of government organisations, financial companies, cellular operators, mining and energy companies, among other businesses, large and small, helping to connect India to a limitless future.

LEO Satellite Constellation

Traditional communication satellites are geostationary and have been in orbit for more than 50 years. GEO satellites weigh more than 1000kg and operate 36,000 kilometers above the earth. These satellites remain in a fixed position relative to any position. Despite Earth's orbit, this allows ground-based antennas the ability to point directly at the satellite, in a fixed position.

In contrast, Low Earth Orbit (LEO) satellites are miniaturized, orbiting versions that operate between 500 and 2000 kilometers above Earth's surface and weigh under 500kg. Due to its low orbit, latency is significantly reduced as the satellite is better positioned to quickly receive and transmit data. Unfortunately, this also creates a smaller coverage area so LEO satellites continuously hand off communication signals and traffic across a constellation of satellites. This ensures seamless, wide-scale coverage over a pre-defined geographical area.

In the 1990s, several companies tried to provide global connectivity using LEO constellations. Globalstar, Iridium, Odyssey, and Teledesic had impressive plans. However, all but Iridium scaled back or canceled their intended constellations because of high costs and limited demand. All suffered financial problems. After that experience, many industry analysts and investors remain skeptical about the viability of large LEO constellations.

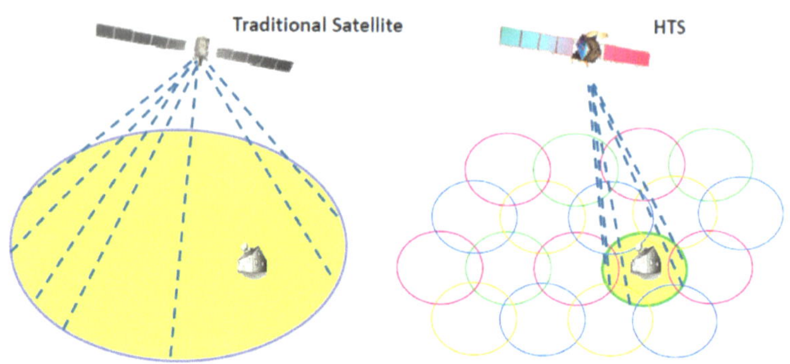

Figure 94: *HTS exclusively uses frequency reuse and multiple spot beam (Image Source: Public Domain)*

But much has changed over the past 20 years. Satellite technology has advanced; demand for bandwidth has soared, with no slowdown in sight; and companies have developed creative business models to generate profits from connectivity.

Recent LEO momentum has been revived by the aggressive launch campaign of SpaceX's Starlink LEO constellation, providing satellite Internet access coverage to 40 countries. It also aims for global mobile phone service after 2023. SpaceX started launching Starlink satellites in 2019. As of September 2022, Starlink consists of over 3,000 mass-produced small satellites in low Earth orbit (LEO), which communicate with designated ground transceivers. In total, nearly 12,000 satellites are planned to be deployed, with a possible later extension to 42,000. Starlink provides internet access to over 500,000 subscribers.

OneWeb is another company that aims to build broadband satellite Internet services. India's Bharti Global, Eutelsat and the Govt. of UK are the company's largest shareholders. Services to begin with Arctic region including Alaska, Canada, and the

UK. OneWeb has signed a pact with New Space India Limited (NSIL), the commercial arm of ISRO to use PSLV & GSLV for satellite launch.

On Oct. 23, 2022 ISRO launched 36 OneWeb satellites in 1st commercial launch for LVM-3. ISRO launched another 36 OneWeb satellites in their second commercial launch of LMV-3 on March 26, 2023. With this, OneWeb's total in-orbit constellation became 218 satellites. These would be a part of OneWeb's 648 LEO satellite fleet that will look to deliver high-speed, low-latency global connectivity in the coming years..

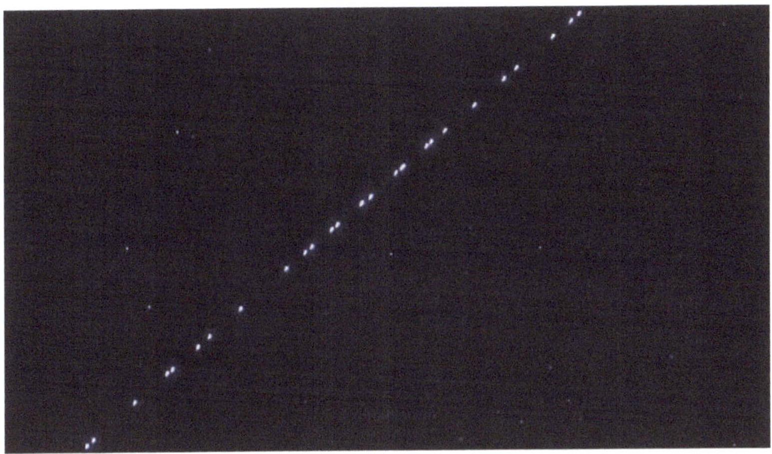

*Figure 95: Starlink Satellites In Night Sky Over Udupi, India
(Image taken by Nihal Amin)*

Software-Defined Satellite

Unlike traditional comm. satellite, software-defined satellite be reprogrammed in orbit to change satellite coverage, operation frequency, bandwidth, and transmission power as per the user demanded.

EUTELSAT QUANTUM heralds a new era of commercial satellite service, where government organizations will be able to actively define and shape the performance and reach from a satellite, as needed. It is the first commercial Ku-band satellite to have a fully flexible payload that can be remotely configured by software from a user's premises. It can be controlled not only by Eutelsat, as the satellite operator, but also by the client, who can control their payload and implement operational scenarios according to their mission requirements.

Figure 96: Eutelsat Quantum satellite
(Image Source: ESA, Public Domain)

Satellite Based Secure Quantum Encrypted Communication

Today's world is ever more interconnected, but these connections are vulnerable to cyberattacks. An ultra-secure telecommunications satellite that uses the unbreakable laws of physics can be used to keep information private. Such satellites employs the laws of quantum mechanics to keep secure the exchange of sensitive information between multiple parties.

Quantum key distribution (QKD) is the use of laser beams to transmit cryptographic keys securely using photons. These photons are coded in binary ones and zeroes and are later picked by the receiving equipment. The Photon has a unique property, any attempt to read and copy a photon will be detectable by the receiver. This makes it possible to transmit keys without getting hacked.

Quantum states such as photons are very fragile, on ground they are lost very quickly due to ground noise while transferring. Satellites can be used for long distances transmission of photon.

In 2016, China launched the satellite 'Micius', which is solely dedicated to quantum information science. In 2017, the concerned Chinese team, along with a group of researchers in Austria, was able to employ the satellite to perform the world's first quantum-encrypted virtual teleconference between Beijing and Vienna.

On 27 January 2022, scientists from Space Applications Centre (SAC) and Physical Research Laboratory (PRL) have jointly demonstrated real time Quantum Key Distribution (QKD) over 300m atmospheric channel along with quantum-secure text, image transmission and quantum-assisted two-way video calling.

With this development, ISRO is getting ready for satellite based demonstrations of fundamental quantum mechanics experiments as well as quantum communication for future-proof data security.

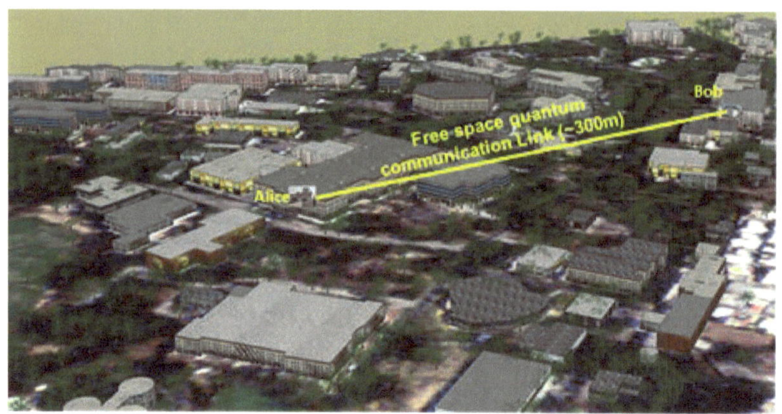

Figure 97: SAC, ISRO & PRL demonstrates entanglement based quantum communication over 300m free space along with real time cryptographic applications (Image: ISRO.gov.in)

32

Communication Faster than Light

Faster-than-light travel is not just science fiction; it's a way to reach out and touch the stars. It's a way to make the universe accessible to humanity.

- Michio Kaku

With our current technology, we can send messages at nearly the speed of light. The special theory of relativity implies that only particles with zero rest mass (i.e., photons) may travel at the speed of light, and that nothing may travel faster.

Now, Alpha Centauri, the nearest star system to Earth, is 4.37 light-years away. Meaning if humans were to ever travel there it would take over 8 years to send a text message and get a response. This means humanity cannot go into the stars without a means of communicating faster than the speed of light. Good thing we already have some pretty good ideas as to how. For that we rely on Quantum entanglement. When two particles are linked together no matter their separation from one another, they still are able to share information with each other immediately.

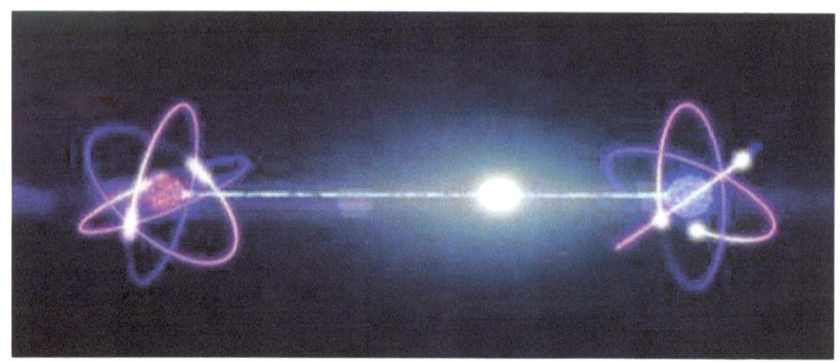

Figure 98: Quantum-Entanglement, Representing Image.
Source: https://tinyurl.com/quantum-entanglement-hg

So how fast is "immediately"? According to research by Prof. Juan Yin and colleagues at the University of Science and Technology of China in Shanghai, the lower limit is around 3-trillion meters per second – or four orders of magnitude faster than light around 3-trillion meters per second – or 10,000 times faster than light [26]

At that speed, you could send a message to Alpha Centauri in just under 3 hours. Not too bad.

Research continues on this subject and some physicists believe that faster-than-light communication might be possible one day.

33

Communication in Distant Future

The laws of physics are not as they seem. Who knows what the future holds for communication? Maybe we will find ways to defy our current limitations.

- Richard Feynman

Communication in Distant Future: Through the Eyes of the Futurologists

Futurologist says Humanity will abandon speech in the future. *"One day, I believe we'll be able to send full rich thoughts to each other directly using technology"*, Mark Zuckerberg said during a June 2015 interview. *"You'll just be able to think of something and your friends will immediately be able to experience it too if you'd like."*

In the distant future, we may be able to communicate by sending our thoughts through a network directly into someone else's brain. We're decades away from such technology, but scientists are working on creating brain-computer interfaces that allow people to transmit thoughts directly to a computer. Perhaps 50 years from now we'll all use an electronic version of telepathy.

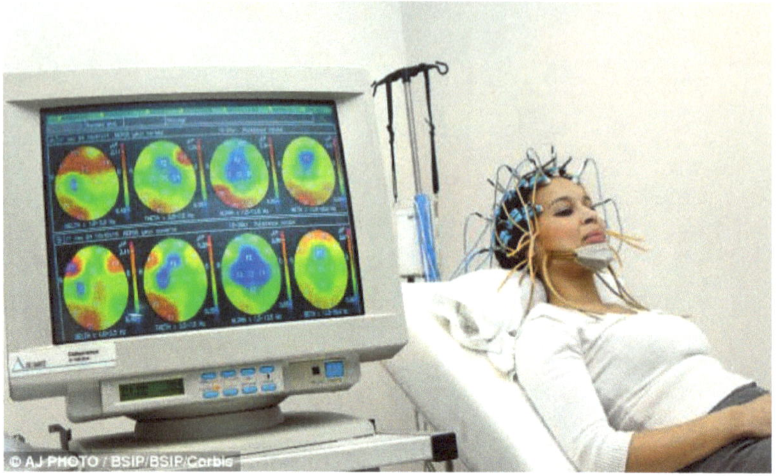

Figure 99: Mind reading Representing
Image Credit: A J PHOTOS/BSIP/BSIP/Corbis

Futurologist Dr. Ian Pearson, whose job is all about forecasting the future, said: *"By 2050, we'll be communicating through a form of telepathy from thought recognition technology. Electronic jewellery which can detect my thoughts may be able to communicate with someone else's earpiece and relay my thoughts to them."*

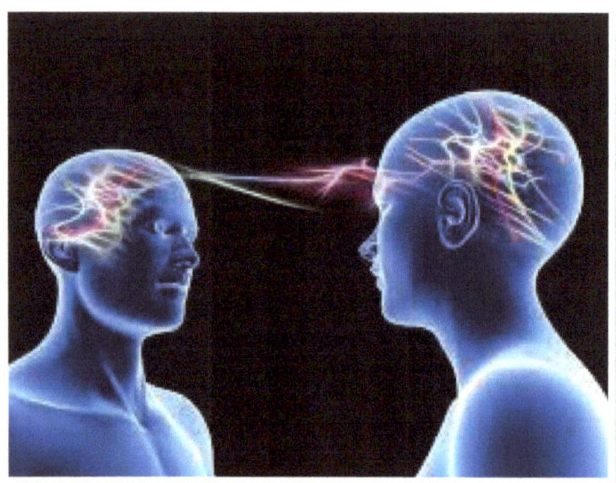

Figure 100: Image showing brain to brain communication
Image credit: ALFRED PASIEKA / SCIENCE PHOTO LIBRARY

Communicating is inherent to human beings, who for millennia have been developing technological solutions to overcome the great obstacle of communication: distance. The smoke signals and Drum languages devised by our ancestors began to address this problem, which a chain of inventions from the telegraph to the satellite have finally left solved... or almost solved. Brain waves and augmented reality promise to be the first step towards the next technological leap. What is the next barrier that human communication will overcome?

References

1. "Harnessing space technology for societal benefits – A brief history of SAC" by SAC/ISRO

2. "Satellite Communication in India – Perspectives on Vision to Reality" by SAC/ISRO

3. "Dr. Abdul Kalam's lesson from failure of ISRO's SLV-3" by Saurabh Tiwari, Linkedin, Sep 9, 2019

4. "Yash Pal: A Life in Science" by Biman Basu

5. "Making of a Satellite Centre" by Prem Shanker Goel

6. "Space and Beyond - Professional Voyage of Kasturirangan" by B. N. Suresh

7. "ISRO: A Personal History" by R. Aravamudan

8. "Early history of wireless communications" by Animesh Maitra, Souvenir of 33rd Annual Convention of Radio Physics and Electronics Association, University of Calcutta

9. "The birth of the Internet in India" by Shauvik Ghosh, Ashish K. Mishra, Mint 30 Jun 2015

10. "The Story Behind Morse Code (and Why It Matters to You)" by Daniel Jacobs, Linkedin.com, 30/4/2016

11. The First "Hello!": Thomas Edison, the Phonograph and the Telephone – Part 2 By Allen Koenigsberg

12. Why Titanic's first call for help wasn't an SOS signal by Yerin Blakemore, National Geography, May 28, 2020

13. Edison vs. Tesla: The Battle over Their Last Invention by Joel Martin

14. When Thomas Edison Tried Besting Nikola Tesla by Building a "Spirit Phone" By Michele Debczak, Mental Floss, Oct 25, 2019

15. The postman of Kusthia by Touhidi Hasan, Prothom Alo Bangla 24 Mar 2017

16. 8 Interesting Facts About Thomas Edison, Asia Pacific Economics Blog, Oct 29, 2014

17. Indian Telegraph by John H. Lienhard, University of Houston

18. Last telegram message sent to Rahul Gandhi, The Hindu, Businessline Published on July 15, 2013

19. Listen to the father of Indian Telecom by Kamla Bhatt, Mint, 26 Jan 2009

20. Sir J.C. Bose diode detector received Marconi's first transatlantic wireless signal of December 1901 (the "Italian Navy Coherer" Scandal Revisited) by P.K. Bondyopadhyay, Proceedings of the IEEE, Volume: 86, Issue: 1, January 1998.

21. The Story of Aryabhata, U.R. RAO, From Fishing Hamlet to Red Planet, Edited by P.V. Manoranjan Rao

22. Biographies Cher Ami By Ellora Larsen, National Museum Of The United States Army

23. 5 Carrier Pigeon Stories" 15 May 2012. HowStuffWorks.com

24. https://en.wikipedia.org/wiki/Drum

25. Delivered by pigeon post in Cuttack by Satyasundar Barik, The Hindu, May 05, 2018

26. Chinese Physicists Measure Speed of Quantum Entanglement by Joshua Filmer, https://futurism.com/

27. APPLE in Retrospect, by R.M. VASAGAM From Fishing Hamlet to Red Planet, Edited by P.V. Manoranjan Rao

28. "The INSAT-2 Story" by P. RAMACHANDRAN From "Fishing Hamlet to Red Planet", Edited by P.V. Manoranjan Rao

29. "20 years of the mobile phone in India: One call that started it all" by Umang Das, Economic Times, August 1, 2015

From Smoke Signals to Satellites

The fascinating story of Communication

D.K. Das, Former Director & Distinguished Scientist, SAC, ISRO
P. S. Bharadhwaj, Former Senior Scientist, SAC, ISRO

In this book, we delve into the fascinating history of communication and explore how it has shaped the world we live in today. From the early days of smoke signals and carrier pigeons to the advanced satellite communications of today, we will follow the stories of the pioneers who pushed the boundaries of what was possible and changed the way we communicate forever.

Through vivid descriptions and compelling human narratives, we will uncover the key moments in the history of communication and the people who made them happen. We will also delve into the rich history of modern telecoms in India, including the inspiring story of ISRO and the Indian space program.

This book is not just about the technical details

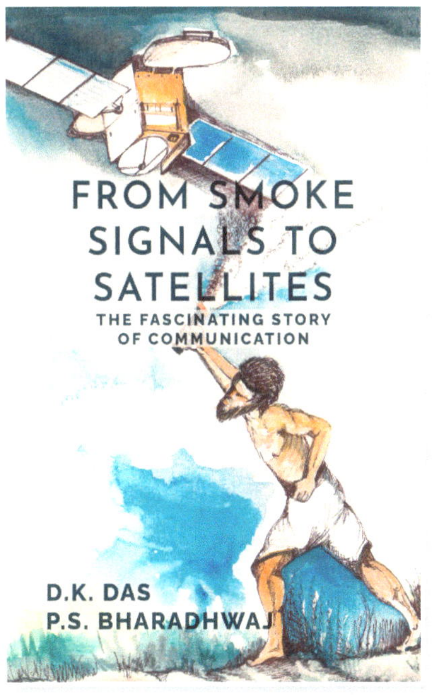

of communication innovations, but also about the human impact they have had on our world. So join us on a journey through time as we explore the evolution of communication and the people who made it possible.

Here are some titbits of interesting information from the book:

- A postman inspired Tagore to compose the National Anthem of Bangladesh.
- The tragic story behind the Morse code & Telegraphy.
- One famous inventor proposed to his wife in Morse code.
- Who sent the world's last telegram?
- Why Do We Say Hello When Answering The Phone?
- Did Titanic receive any iceberg warning?
- Did Thomas Edison build a ghost phone?
- Who made the first mobile call in India?
- Film actor Shammi Kapoor was one of the first Internet users in India.
- How did Sarabhai obtain the land in Jodhpur Tekra, Ahmedabad to start the Indian Space Program?
- Prof. Satish Dhawan drew a salary of Rs. one as the Chairman of ISRO.
- Balloon communication was attempted for live telecast of SLV-3 launch.
- Minutes before the take-off of SLV-3, a technician had to climb the launch tower for last minute adjustments.
- Apple satellite was transported on a bullock cart.
- Prime Minister Rajiv Gandhi talked to US President Ronald Reagan for one component of INSAT-2A